KV-694-296

Telecommunication Systems Design

Volume 1

Transmission Systems

Telecommunication Systems Design

Volume 1 Transmission Systems

by

M. T. HILLS B.Sc., Ph.D., C. Eng., M.I.E.E.

B. G. EVANS B.Sc., Ph.D., C. Eng., M.I.E.E.

Department of Electrical Engineering Science
University of Essex

London

GEORGE ALLEN & UNWIN

Boston Sydney

First published in 1973
Second impression 1979

GEORGE ALLEN & UNWIN LTD
40 Museum Street, London WC1A 1LU

ISBN 0 04 621018 0

Printed in Great Britain by
Biddles Ltd, Guildford, Surrey

Preface

by J. H. H. MERRIMAN
C.B., O.B.E., M.Sc., A. Inst. P., C.Eng., F.I.E.E.
Board Member for Technology and Senior Director: Development – British Post Office

Engineers do not need to be reminded of the technical complexity and subtlety of telecommunications equipment. But they do need to be reminded of the significance of two major changes in attitude to telecommunications system design that are now happening. The first is that of the growing importance of the system, as a whole. No longer can significant elements of any system be considered in isolation. Their interaction and interdependence (in economic as well as in engineering terms) are in some ways more significant than their individual attributes. The second major change is the crucial significance of the advice, recommendations, and standards of the Consultative Committees of the International Telecommunications Union. Any telecommunications designer who neglects these does so at considerable risk.

It is therefore good to see in the scope of this volume and of the volume which is to follow it that the authors pay so much attention to fundamentals. For at a time of great technological change, the fundamentals that lie behind design principles cannot be over-emphasized.

Contents

Introduction

This book has been designed to provide an introduction to the design problems of transmission systems which are to be used as component parts of a large telecommunication system. The problems considered are the basic principles of interconnected transmission systems and how in practice one obtains specifications for the transmission system. A large part of the book deals with the national and international telephone system since this is the most important example of a large telecommunications system, and nearly all other forms of telecommunication rely upon the components of the telephone network in some form or another.

The approach of the book has been to isolate the basic principles involved in the design and where necessary to discuss the techniques employed. In such a limited text one can only outline many of the problems and considerable reference is made to the literature for further reading, mainly from the British and North American sources, together with the publications of the International Telecommunications Union.

The aim of the book is to provide the student who has had an introduction to the basic principles of telecommunications with an insight into the real life design problems of systems, and the ability to read the literature to find detailed solutions and techniques. The book should be of value to any undergraduate or post-graduate course which goes in depth into telecommunication systems, and it should be invaluable for anybody who is working in the field of telecommunications, either for an operating administration or for a manufacturing industry.

An explanation of the problem is given in Chapter 1. Chapter 2 is devoted to an introduction to the basic principles of telephony and, in particular, the effects and limitations of amplification. Chapter 3 gives an account of how the existing national telephone networks are planned with particular reference to the British and North American systems, and how these are then combined with others to produce an international system. The contents of these chapters are relevant whatever the transmission technique used. Chapter 4 gives the basic principles of frequency division multiplex systems and this too is independent of the transmission media used.

The next few chapters discuss the particular transmission techniques and the effect they have on the design problem. Chapter 5 gives an overview of the different propagation mechanisms that may be used for communication and in Chapter 6 some of the problems of using small capacity radio systems are discussed. Chapters 7 and 8 deal with the specific techniques of terrestrial microwave and satellite systems. Chapter 9 outlines how the transmission system so far considered may be used or adapted for programme and television uses.

The next three chapters deal with digital techniques and transmission. Chapter 10 describes some of the relevant properties of a pulse code modulation system and how they may be used for speech and other signals. Telegraphy and how it uses the existing transmission system is covered in Chapter 11, whilst Chapter 12 covers the area of data transmission.

Finally, in Chapter 13 there is a brief review of some possible transmission systems that may be used in the future and their implications on the system design problem.

The authorship of the chapters is B. G. Evans for 5, 7, 8 and 13, and the rest by M. T. Hills.

This book results from courses given by us in the M.Sc. course in Telecommunication System Design in the Department of Electrical Engineering Science at the University of Essex for the past three years. We are deeply indebted to the British Post Office who made this course possible by the aid of very generous grants to the University for the purpose. The content of the book is the result of considerable interaction with our students who provided the stimulus and the criticism, without which the book would never have appeared.

We are also very grateful to Professor K. W. Cattermole for his detailed criticisms of the work and we would like to acknowledge the many vital discussions we have had with members of the British Post Office, in particular D. L. Richards, S. Munday and A. Jefferies. The responsibility for errors and obscurities, however, remains with us.

Finally our thanks are due to Mrs C. Newman who over the years has learnt to read our writing so effectively and has typed the many versions of the manuscript.

M. T. Hills
B. G. Evans

Chapter 1

System design problem

1.1 Introduction

This book describes the overall design problems and some of the solutions associated with the design of large-scale switched telecommunications systems. The major concern will be with the national and international networks that are, or will be, used for speech, video, telegraph or data transmission. Stated in simple terms, the problem is to match the transmission channel to the signal in an economic manner. Since transmission channels are non-ideal, it is necessary to specify the maximum limits of the allowable degradations on the individual links that may be used in a switched system such that acceptable communication is obtained at the lowest cost. Where links are international it is also necessary to specify a suitable set of standards in order to permit interconnection of the networks in different countries.

The aim of this book is to show how the allowable degradations are found and how they may be allocated to the different links in an economic manner. There will be considerable reference to the technology used in the different transmission techniques where this has relevance to the overall planning problem. The standardisation necessary to provide for international communication is performed under the auspices of the International Telecommunication Union (ITU). Although this is a purely advisory body its recommendations are very widely applied for purely internal links as well as the international links. The ITU have been in existence for well over 100 years[1] and they have made possible the fantastic global communications system that is in existence today. They work nowadays through two main committees.

1. CCIR (Comité Consultatif International de Radiocommunications– International Consultative Radio Committee). This was set up in 1927 with the duty to '. . . study technical and operating questions relating specifically to radio communications and issue recommendations on them'.
2. CCITT (Comité Consultatif de Téléphone et de Télégraphie– International Consultative Committee for Telephony and Telegraphy). This committee was formed in 1956 when the CCIF (Telephone Committee) and the CCITT (Telegraph Committee) amalgamated. Their

duties include the study of tariff questions as well as technical and operating questions relating to telegraphy and telephony. Their studies include data and video transmission.

These committees are made up of representatives from telephone administrations, scientific and industrial organisations etc., and they meet in plenary sessions every three or four years. Between the sessions various working parties consider particular questions and produce recommendations for system standards, operating procedures, tariff structures etc., and these recommendations are discussed in the plenary sessions. The results are issued as volumes of recommendations and in the case of the CCITT volumes they are known by their colour. Recent meetings and their recommendations have been:

CCIR:
1966 Oslo XIth Plenary assembly
1970 New Delhi XIIth Plenary assembly

CCITT:
1960 New Delhi Red Book series
1964 Geneva Blue Book series
1968 Mar del Plata White Book series

In view of the importance of these bodies, there will be frequent mention of the specific recommendations adopted in the book.

Although our only interest here is with communications with the aid of electrical signals, it is important to realise their interdependence with other forms of communications. In many cases there is a realistic choice between physical and electrical communication. For instance, some banks can choose whether it is more economic to transfer paper tape by courier to a central processor or else use a data link. Also a man can decide whether to use the telephone, write a letter, go personally or use some form of vision telephone, if available. The interesting point is that the electrical possibility is almost always preferable provided that it is sufficiently economic, although there are some obvious counter examples.

Let us first examine the basic electrical characteristics of the different types of information that our systems are required to deal with.

1.2 Types of information and their characteristics

When dealing with an electrical signal, the most obvious parameter that can be specified is its bandwidth if it is an analogue signal, or its bit-rate if it is a digital signal of some form. The major types of information may be placed in one of the following categories with the approximate bandwidth as shown:

(a) Telegraph or teleprinter 50 bits/s
(b) Telephone quality speech 300-3400 Hz

(c)	Music channels for broadcast systems	50 Hz-15 kHz
(d)	Facsimile for newspapers, weather maps, finger prints	e.g. with a 40 kHz bandwidth 1 page of newspaper can be transmitted in 25 min.
(e)	Visionphone applications	From as low as 500 kHz for simple systems
(f)	Closed circuit television	From 1 to 5 MHz depending upon application
(g)	Broadcast television	0-5·5 MHz for 625 line
(h)	Data	Up to 1 Mbit/s for direct computer to computer communication

For the analogue signals it is impossible to give a reasonable specification on theoretical grounds of the channel requirements necessary to transmit them. In the case of telegraph or data signals where the message is well defined, then an objective 'probability of error' may be stated. For signals which are primarily derived from speech or visual sources, the system requirements are very subjective since they have a considerable redundancy and consequently impairments may be introduced without loss of intelligibility.

As an example of this redundancy, Shannon[2] has shown that the total information handling capacity of a channel of bandwidth W Hz and a signal-to-noise power ratio S/N is given by

$$C = W \log_2 \left(1 + \frac{S}{N}\right) \text{bits/s}$$

Thus for a good telephone line with a 3 kHz effective bandwidth and 40 dB (≡ 10 000:1) signal-to-noise power ratio,

$$C = 3000 \times \log_2 10\,001 \simeq 40\,000 \text{ bits/s}$$

However, how much information can one actually convey in one second? For example, one could say 3 decimal digits in one second giving a repertoire of 1000 messages, i.e. an information rate of about 10 bits/s ($2^{10} = 1024$)–a total efficiency of 0·025 per cent. The inefficiency of television signals is even more astronomic. Hence the quest in many research establishments throughout the world is to find the meaningful content of speech and vision so that some alternative form of coding may be found which gives a degree of bandwidth compression. To date any practical bandwidth compression system has involved such a large investment in equipment that these systems are totally uneconomic for any except the most expensive communication links such as satellite links. Since the trend is for the cost of bandwidth to be reducing then the necessity for such equipment is reducing also. The most cost-effective form of bandwidth compression, both speech and visual, is simple bandwidth limitation, i.e. a low-pass filter.

Volume of a speech signal

The other main characteristic of an electric signal is its power and dynamic range. This is most difficult to define in the case of speech signals since it is a complex non-periodic function, whereas although data and video signals are complex, they do have well defined limits of variation of the signal levels.

Measures such as r.m.s., peak or average, are more relevant to periodic signals than to speech, but for planning and operational purposes it is necessary to have a single quantity which adequately characterises the signal and can be easily measured. A convenient measure of speech volume, used by the British Post Office, is the *mean power whilst active,* together with the proportion of time a talker is active[3]. By experience it has been found that a suitable definition of silence is those periods of at least 350 ms for which the short-term mean power (averaged over 20 ms which is the length of a syllabically significant segment) is at least 15 dB below the mean power whilst active. The unit of measurement is the dBm, which is dB relative to 1mW.

An alternative unit which is used in North America is the volume unit (VU). For this unit the primary functions required are stated to be:

(a) Measure signal magnitude in such a manner that the user can check for overload and distortion.
(b) Check transmission gain and loss.
(c) Indicate relative loudness of signal when converted to sound.

A peak reading instrument is not adequate for this purpose since the effect of delay distortion which leaves the received sound unaffected can affect the peak values of the waveform. For this reason a r.m.s. meter is chosen which in effect integrates the signal over a short period of about that of a syllable. In actual use some practice is necessary since a mental 'average' of the deflection must be taken, ignoring the occasional high-valued peaks. This can then be related to the dynamic range of the system[4]. The definition of a volume unit is thus the meter reading taken by somebody who is expert at taking these readings[5]. This type of unit is familiar to tape recorder users with a meter indicator.

The VU indication is in effect a measure of mean power whilst active and by means of empirical measurements it is found that the mean power is given by

$$P_A = \text{VU} - 1{\cdot}4 \text{ dBm}$$

Hence the long-term mean power is given by

$$P_M = \text{VU} - 1{\cdot}4 + 10 \log \tau \text{ dBm}$$

where τ is the activity ratio. For $\tau = 0{\cdot}25$, $P_M = \text{VU} - 7{\cdot}4$ dBm.

This type of empirical definition usually comes as a shock to people who have been trained to regard engineering as an exact science but, because speech is such a complex and ill-understood process, the use of these empirical definitions, which have been found by years of experience to be useful in design, is inescapable.

1.3 Review of channel characteristics and their degradations

The main transmission media available for telecommunications and their basic characteristics are:

(a)	Copper conductors:	Parallel on telegraph poles or twisted pair. Can be used up to several hundred kilohertz if adequate amplification and balancing is provided. Aluminium conductors are now being used for the local network as they show an economic advantage.
(b)	Coaxial cables:	Usable from about 50 kHz to hundreds of megahertz.
(c)	Microwave:	Either terrestrial or satellite in the frequency ranges 2, 4, 6 and 11 GHz
(d)	Tropospheric scatter:	Very high-power single hop over the horizon using frequency range 0.8 to 5 GHz. Distances 100–600 km. Bad fading experienced.
(e)	H.f. radio:	Point-to-point around the world in range 3-30 MHz. Frequencies of use vary with time of day, sun-spot activity etc. Bad fading experienced.
(f)	V.h.f. radio:	In range 150 to 180 MHz–usable to just over the horizon for small scale point-to-point and mobile operation.
(g)	U.h.f. radio:	Of increasing use for mobile operation
(h)	Other media:	Research is under way on: Microwave waveguides Fibre optics Laser beams and the potentiality of these will be discussed in Chapter 13.

In addition to the bandwidth limitations there are many other channel degradations which can be grouped under the following headings:

1. *Those which affect one-way communication*
   (a) Attenuation–loss at some nominal frequency and its variation with time.
   (b) Attenuation/frequency distortion–change of attenuation with frequency.

(c) Amplitude distortion–change of attenuation with amplitude (this may either be with the instantaneous amplitude or some function of the previous signal).

(d) Phase or delay distortion–change of propagation delay with frequency.

(e) Dynamic range limitation– (i) peak power
(ii) mean power
(iii) minimum power (e.g. on a quantised system such as p.c.m., see Chapter 10)

(f) Noise–power, bandwidth and statistics of circuit noise and, in the case of speech, the background room noise.

(g) Cross-talk from other channels.

(h) Sidetone and echo. (See next chapter)

2. *Those which affect two-way communication*
(i) Delay or propagation time.

(j) Effects of voice-operated devices such as echo suppressors etc.

These degradations are all defined in the frequency and/or time domain and are in principle easy to measure, but they bear little relation to the parameters of analogue signals. In order to find out the maximum impairments that a channel can introduce to a signal it is necessary to resort to experiment. These degradations depend considerably upon the application, and the allowable magnitudes may well change with time as users' expectations increase. The first voice contacts from Europe with the USA or Africa were barely intelligible, but were hailed as marvellous, but nowadays people are increasingly expecting as good quality for these calls as on inland connections.

The specification of the limits of the degradations are of two main types:

(a) Extreme limits beyond which no satisfactory system must go.
(b) Tradable limits which can be varied according to the degree of satisfaction required.

A large proportion of what follows is an account of how these limits are found, and, where applicable, how they may be traded for one another.

Noise measurement

The degradation that is most difficult to measure is that of noise since it is a complex waveform and thus poses similar problems to that of speech volume measurement. The quantity of interest is not the absolute power of the noise but a measure of its interference to a listener or viewer, or to the reception of data. The desirable characteristics of a measurement are that[6]

(a) different noises which are judged to be equally interfering are given equal numerical magnitude
(b) the sum of the measurements for separate noises should be a good indicator of the interfering effect of the noises combined.

Since the interference effect of a single-tone noise is frequency dependent then some form of frequency weighting is necessary to equalise the effects. These weightings may be found by means of subjective experiments and various cures have been standardised for specific applications. The weighting adopted for telephone speech is shown in Figure 1.1. In practice it has been found that

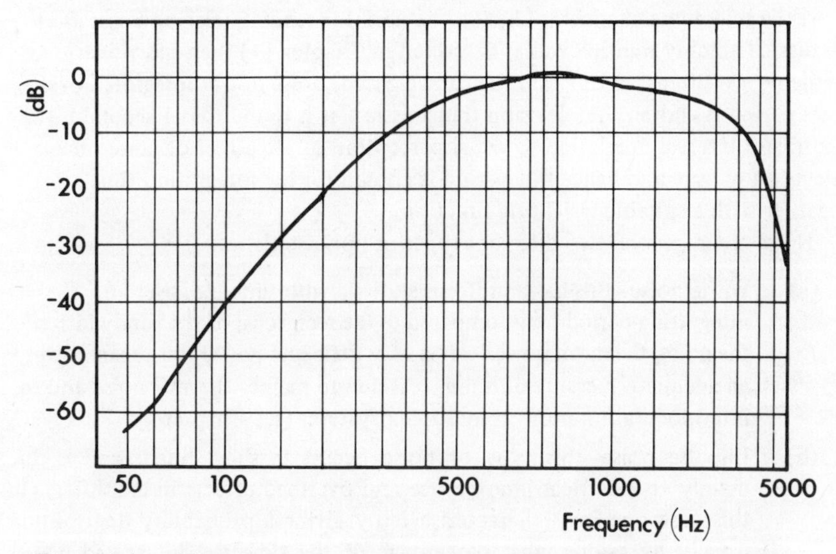

Figure 1.1 Characteristic curve of the psophometer filter network used for measurements at the terminals of a commercial trunk telephone circuit (from CCITT *White Book*, Volume V, Recommendation P53)

adding components of noise together on an r.m.s. basis (i.e. power addition) provides an adequate representation of their combined effect. The duration of the noise is also important since, in the case of speech for instance, noise bursts of less than 200 ms length have less interference effect than their absolute power would indicate. This factor may be included by building a suitable transient response into the indicating meter such that short bursts have less effect. The meter is then read in a similar fashion to a volume meter. The components necessary for a noise meter are shown in Figure 1.2[7] and such devices are called *psophometers.*

In the field of transmission system design it has been found that longer term averages of the noise power are a more useful indicator of channel behaviour. In the case of telephone speech it has been found that integrating the weighted

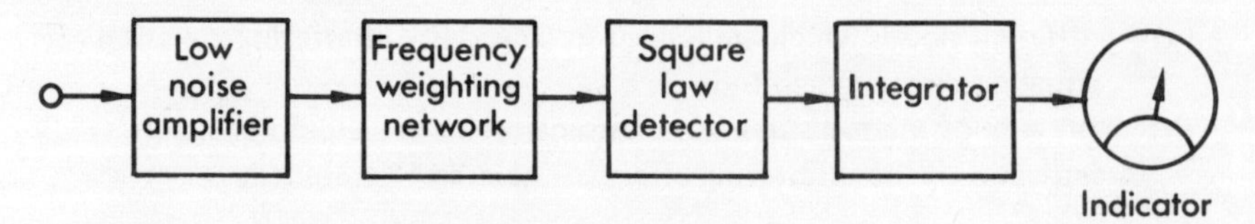

Figure 1.2 Components of a psophometer

noise power over a one minute period provides a good measure of its annoying effect [8]. This time is chosen to be comparable to the assumed average length of a telephone call of approximately 3 minutes.

If the telephone channel is likely to be used for sending telegraph signals at 50 bits/s (suitably multiplexed as described in Chapter 11) then individual elements last 20 ms. In this case the (unweighted) noise power integrated over a 5 ms period is chosen. In television transmission, it is found that 1 second is the significant interval in relation to visual perception of the effect of noise on a television picture and hence this period is chosen for the integration time together with a suitable weighting function.

There are in general two different types of noise statistics:

(a) White noise—this has a uniform statistic with time, i.e. over an integrating period long compared to the reciprocal of the bandwidth of the noise, the mean power varies very little and hence this mean power is an adequate measure of noise. It is due to mainly thermal noise and to intermodulation noise in multiplex systems (see Chapter 4).

(b) Impulse noise—this type of noise occurs in short bursts and is due mainly to electrical interference and overload effects in amplifiers. In this case one may characterise it by giving a probability distribution usually by saying what percentage of the time the integrated power exceeds a given power level.

In addition there are other types of noise such as quantisation noise in p.c.m. systems which is a function of the signal level. This type will be treated separately in Chapter 10.

In most cases of interest it is possible to design a system which has a predetermined level of white noise but impulse noise can only be found by experimentation, since it is predominantly due to interference of various sorts.

Units of noise power

The integrated noise power is usually expressed in terms of pW (10^{-12} W) and if the noise is frequency weighted the units are often expressed as pWp (p for psophometrically weighted). In Europe the powers are also expressed in terms of dB relative to 1 mW, i.e. dBm or dBmp if weighted. For instance, for telephone quality speech the standard frequency weighting function is effectively 3·1 kHz wide (300-3400 Hz approximately). If the power of a band of white noise 0 to 4 kHz wide is measured with and without this standard weighting, then the weighted power will be 3·6 dB less than the unweighted power, i.e. if the power of a 0 to 4 kHz band of white noise is x dBm then this will give a power of $(x - 3{\cdot}6)$ dBmp when weighted.

If the noise power is measured (or referred to) the zero level reference point of the system (which is discussed in the next chapter) then the units are given as pW0, pW0p, dBm0 and dBm0p.

In North America a series of different units have been developed for noise measurement. Instead of referring the measured noise power to 1 mW they refer it to 1pW of a 1000 Hz signal, in order to make all measurements positive, i.e. they refer to –90 dBm rather than 0 dBm. This unit they called dBrn (the n standing for reference noise). With a frequency weighting called the C-message weighting (which is effectively the same as the CCITT standard) the unit was dBrnc and this is in common use today. An earlier unit was the dBa which had a different frequency weighting and sensitivity. The difference between dBa and dBrnc depends upon the noise characteristic, but on average dBrnc is 6 dB higher. An approximate translation table between the European and the old and new North American units is thus, to convert from *A* to *B*:

Table 1.1

A \ *B*	dBmp	dBrnc	dBa
dBmp		+90	+84
dBrnc	–90		–6
dBa	–84	+6	

i.e. 26 dBrnc is the same as –64 dBmp and approximately the same as 20 dBa.

1.4 Specification of allowable degradation for speech transmission

The ideal requirement for a speech transmission system may be stated as: 'The ability to converse with zero effort'.

This is a very nebulous statement and a more positive form is required in order to find suitable design parameters. A statement suitable for a civilian speech system is to find the parameters such that (for instance) '95 per cent of the (normal) users would consider the system satisfactory'.

The job of the system designer is then to provide just these conditions at minimum cost and also to ensure that the unlucky 5 per cent of the calls that fall below the limit of satisfactory are not always the same users. The choice of the 95 per cent figure is a management decision but the definition of satisfactory is the engineers.

In a more specialised application, such as air traffic control, less stringent requirements may be made and these will usually relate to intelligibility.

The parameters of a transmission channel that will yield a suitable percentage of satisfied speakers have been found as a result of years of accumulated experimental effort throughout the World. As new transmission techniques (such as p.c.m.) and new handset designs are introduced into the system, these experiments have to be repeated.

These experiments are usually performed with the aid of a representative standard communication link together with typical terminal equipment. Arrangements are made to introduce into the link specific degradations such as loss, noise etc. The link is assessed for satisfactory performance, and since this assessment is based on the opinions of human subjects, the procedures are often long and complex. There are three main types of assessment tests that may be performed[9,10].

(a) *Conversational tests.* This is the most direct method of assessment whereby pairs of subjects are introduced into a realistic telephone environment and perform some task involving mutual conversation[11]. This task, for instance, could be some puzzle which involves the interchange of names and numerals. On completion of the task the subjects express their subjective opinion of the link on a five point scale.

4 excellent
3 good
2 fair
1 poor
0 bad

For a given set of circuit conditions and a large number of pairs of subjects it is then possible to find the 'mean opinion score' for that set of conditions. If the opinions 'poor' or 'bad' are taken to be unsatisfactory then it is possible to convert the mean opinion score to a percentage of unsatisfactory calls. Unfortunately it is not possible to predict the increase in unsatisfactory calls when a combination of degradations are present from a knowledge of their effect by themselves. For instance, the addition of a certain level of received noise on a link with a particular loss may increase the percentage of unsatisfactory calls from 0·7 per cent to 1·0 per cent. If the loss is increased by 12 dB then the addition of the same level of received noise is found to increase the percentage of unsatisfactory calls from 16 per cent to 28 per cent. This implies that the measurements of mean opinion score must be made over all combinations of all degradations, and since each measurement takes a considerable time, this is an expensive and time-consuming technique.

(b) *Listening tests.* A simple and faster assessment technique is to use trained operators who make assessments on the basis of listening only. In this case it is necessary to have some more accurate measurement procedure than mean opinion score, and the technique that has been found to be very useful is the use of a standard test system, which has well defined sensitivities and frequency response. This system contains an attenuator and the listener adjusts this attenuator until she considers that the received speech is equally loud to that of the test speech link. The setting of the attenuator then gives a measure of the relative efficiency (on a loudness basis) of the test and standard system, and is referred to as the *reference equivalent*. This is measured in dB and is positive if the test system is *less* sensitive than the standard, i.e. if the attenuator is adjusted to 20 dB in order to produce equal loudness of speech through the test and standard systems then the reference equivalent of the test system is said to be 20 dB r.e. The history and development of the standard circuit used is described in Chapter 3.

The actual sensitivity of the standard system is arbitrary and is chosen primarily to make it more sensitive than any test system that it is likely to be used with. This means that it is never necessary to use 'negative' attenuator settings. To give some idea of the actual magnitude of the sensitivities, some figures are useful. If the standard system is adjusted so that it has equal loudness with a 1 m air path between a speaker's mouth and a listener's ear, then the reference equivalent is found to be 33 dB r.e. The preferred value of attenuation for a telephone system using a linear microphone is found to be about 12 dB r.e. (i.e. 21 dB louder than hearing over a 1 m air path). If a carbon microphone is used this produces distortion, and to produce equivalent loudness 3 dB r.e. is found to be the preferred value (i.e. even louder). A maximum reference equivalent of 40 dB r.e. is found to be the limit of satisfactory performance. If the reference equivalents are unequal in either direction, it is found that, within wide limits, the asymmetry by itself does not cause any increase in dissatisfaction, and the mean opinion score rating given to the connection is the same as that which would have been given to a connection in which both directions had the higher reference equivalent.

Reference equivalent measurements may be related to percentage unsatisfactory calls by performing a series of conversational tests and the results obtained are shown in Figure 1.3. With this data it is possible to measure the reference equivalents of a series of circuit conditions and immediately obtain the percentage of unsatisfactory calls for combinations of these conditions. A more detailed discussion of the application of the concept of reference equivalent will be found in Chapter 3.

An alternative use of the standard system is to measure what is called the impairment of a degradation. In a good quality speech system the major factor for satisfactory operation is found to be the loss; hence it is possible to equate other degradations against an equivalent loss. For instance, if a certain degree of bandwidth limitation is introduced into the system, then it is possible to measure the amount of increased gain necessary to restore the received signal to an equally loud level. This increase in gain necessary to compensate for a degradation is called the impairment of the degradation and an example is shown in Figure 1.4. Similar measurements may be taken for the additional noise but,

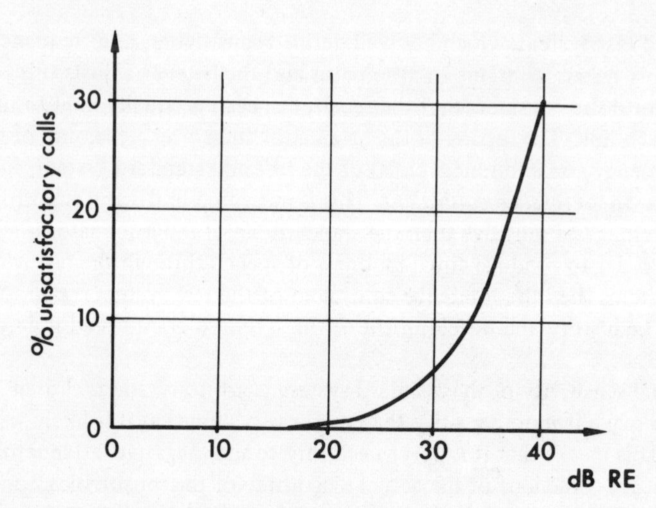

Figure 1.3 Relationship between reference equivalent and percentage of unsatisfactory calls

as mentioned previously, these impairments do not add in a simple manner. However, the overall impairment of a combination of degradations does provide a figure of merit for a link which does not depend upon the arbitrarily defined characteristics of a standard system.

An alternative to assessing a link on the basis of equal loudness to that of a standard system is to use a more objective measure such as articulation

Figure 1.4 Transmission impairment due to limitation of the frequency bandwidth effectively transmitted (from CCITT *White Book*, Volume V, Recommendation P14)

efficiency [12]. In this technique a series of nonsense syllables are sent along the link, and the percentage correctly received is noted. Tests are performed alternatively on the standard system and the test system with different values of attenuation in each system. The difference in these attenuations for 80 per cent articulation efficiency is defined to be the articulation reference equivalent *AEN* (Affaiblissement équivalent pour la nettéte–articulation reference equivalent). This method is not used much nowadays since the standard of telephone connections is such (or is planned to be) that nearly 100 per cent articulation efficiency is assumed and the method is not very sensitive at this level.

(c) *Objective measurements.* It would obviously be preferable if the measures of reference equivalent could be made purely by instruments and without the use of human subjects. In order to achieve this, several problems must be solved.

(i) Suitable artificial ears and artificial mouths must be designed which give accurate, and more important, reproducable results when used with telephones.
(ii) Some definition in electrical terms of what is meant by equal loudness is necessary. [13]

At the present time, neither of these problems have been completely solved satisfactorily, although much research is being performed in this area. For routine measurements on a production line, automated test equipment is available†.

1.5 Financial considerations

The problem posed to an engineer is not how can you make a communication system, but what is the most economic way. It is possible to communicate with almost every point on the earth and near space, but many situations would require inordinate cost penalties. How can one compare cost accurately? One system may require a very high initial investment and then negligible maintenance or power costs such as a satellite, whereas other systems have a lower initial cost but require expensive maintenance.

A common way of comparing the costs of different systems in the telecommunications field is that called '*present value of annual charges*', or in the jargon, the p.v. of a.c.

The basis of such measures is to determine how much money must be invested now at a given rate of interest which will provide for all the capital goods and all annual charges. For instance, to compute the cost of power there is the initial investment in supplies, etc., plus an annual charge for maintenance and electricity bills. If £1 a year is needed for ten years, and the rate of interest is 5 per cent, then £7·722 invested would give this amount. Thus if by spending

† This equipment (type 335) is manufactured by Bruel and Kjaer, Denmark.

£5 more on a circuit the annual power costs are reduced by £1 a year over ten years, then this results in a saving of £7·7, a net saving of £2·7. However, one may not have this extra £5 available at the time and therefore the less economic alternative must be resorted to.

One factor which affects these calculations considerably is the life of the plant or the amortisation period. There is a significant difference between whether a piece of equipment is amortised over ten years or twenty years; at the moment this difference, when applied to submarine cables as opposed to satellites, can tip the balance between the relative costs of the two. Thus, in practice, a mixture of techniques is likely to continue.

There is also the problem of advance plant provision. How much is it worthwhile to instal plant in larger blocks ahead of need, i.e. before they start earning revenue? These problems can be partially solved on a p.v. of a.c. basis.

The whole technique is only a sub-system optimisation and may not be the best. For instance, no charge is made for digging up busy roads and the increased cost of congestion to the community as a whole. If a tax was made for restricting traffic flow then this would change the picture. Also it is difficult to include items such as adaptability etc.

Discounted Cash Flow

Although the p.v. technique gives a relative measure of profitability for different schemes, it does not provide an absolute guide to decision. In recent years the concept of 'return on capital' has been employed in order to provide a better guide to the efficiency of capital expenditure.

Basically this technique involves the construction of a 'cash flow' statement in which each year's net income is recorded, all running expenses and further capital payments are financed out of income and these net sums are discounted at some interest rate r to the start of the project. If the project is to break even, then the following equation must hold:

$$C = \frac{a_1}{1+r} + \frac{a_2}{(1+r)^2} + \ldots \frac{a_n}{(1+r)^n}$$

where C is the initial capital expenditure, a_1 etc. are the net annual incomes and n is the number of years of the project. Suitable allowances have to be made for the residual value of the equipment but no sinking fund or interest payments need be considered since these are included automatically.

If a suitable value of r may be found which satisfies this equation, then this gives the equivalent rate of return on the capital. Computational difficulties can arise if any of the annual net incomes are negative, since the equation then has multiple real positive roots and a suitable choice must be made. In practice this scheme can be reduced to a single go/no-go decision. In this case each annual net income is then discounted by a suitable factor, $(1+r)^{-n}$, and if their sum is less than the initial capital outlay, then the scheme is unacceptable on financial grounds. However, as with p.v. computations, it is difficult to include other considerations such as social value etc.

REFERENCES

1. An interesting history may be found in their centenary publication, *From Semaphore to Satellite*, I.T.U., Geneva, 1964.
2. For example see M. Schwartz, *Information, Transmission Modulation and Noise*, McGraw Hill, 1959, Chapter 1.
3. R. W. Berry, 'Speech-Volume Measurements on Telephone Circuits', *Proc. I.E.E.*, **118**, February 1971, pp. 335-338.
4. Chinn *et al.*, 'A New Standard Volume Indicator and Reference Level', *B.S.T J.* **19**, January 1940, pp. 94-137.
5. The results of an American survey of speech volumes may be found in K. L. M. McAdoo, 'Speech Volumes on Bell System Message Circuits', *B.S.T.J.* **42**, September 1963, pp. 1999-2012.
6. A. J. Aikens and D. A. Lavinski, 'Evaluation of Message Circuit Noise', *B.S.T.J.* **39**, 1960, pp. 879-909.
7. W. T. Cohran and D. A. Lavinski, 'A New Measuring Set for Message Circuit Noise', *B.S.T.J.* **39**, 1960, pp. 911-931.
8. D. Turner, 'Equipment for the Measurement of Some Characteristics of the Noise in Telephoning and Television Channels', *Post Office Electrical Engineers Journal* **55**, 4, January 1963, pp. 231-236.
9. J. Swaffield and D. L. Richards, 'Rating of Speech Links and Performance of Telephone Networks', *Proc. I.E.E.* **106B**, April 1958, pp. 65-76.
10. D. L. Richards and J. Swaffield, 'Assessment of Speech Communication Links', *Proc. I.E.E.* **106B**, April 1958, pp. 77-92.
11. D. L. Richards, 'Transmission Performance Assessment for Telephone Network Planning', *Proc. I.E.E.* **111**, May 1964, pp. 931-940.
12. The technique is described in *CCITT White Book V.*
13. D. L. Richards, 'Loudness ratings of telephone speech paths' *Proc. I.E.E.* **118** March/April 1971, pp. 423-436.

Chapter 2

Basic principles of line communication

2.1 Basic telephony

Nearly all large scale speech communication systems use some form of carbon microphone coupled to a moving iron, or more recently, a rocking armature receiver[1]. The major advantage of the carbon microphone is in its sensitivity and cheapness. It also has a non-linear response which has the beneficial effect of discriminating between speech and room noise. It is not necessary to use any amplifying devices for the majority of connections in a local telephone network (i.e. in the region 10-20 miles, or 15-30 km). However, nowadays the disadvantages of amplifying devices is not as great as it was in the valve era, and work is under way for the development of suitable higher quality microphones which can utilise some form of semiconductor amplification. Of particular interest is the use of moving coil microphones and more recently, the 'electret' microphone[2] which is based on a permanently charged dielectric diaphram. The major problem in both cases is to find systems which are sufficiently robust to be used in the field and whose electronics can withstand the impulses that inevitably occur on transmission lines.

The simplest method of interconnection is that shown in Figure 2.1(a). In practice, if more than one conversation must be taken along the same route, then a form of balanced conductor is necessary to avoid cross-talk and minimise pick-up from nearby power-supply lines etc. These conductors can take the form of either parallel wires on telegraph poles or a twisted pair in an underground cable.

Since the internal impedance of a microphone is much less than that of the line, then greater power transfer into the line may be achieved by the use of a transformer as shown in Figure 2.1(b). The use of local batteries is generally uneconomic (at present) because of the necessity for service and replacement. For this reason, it is preferable to have a large battery placed centrally at the local exchange. Since the voltage across a carbon microphone is proportional to the current flowing through it, then the nearer this central battery is to the instrument the better, and this means that it is better to have a central battery for each instrument located at the nearest telephone exchange. One technique for feeding the current into the line is shown in Figure 2.1(c) where an auto-transformer is used in the instrument to match the impedances. A capacitor is

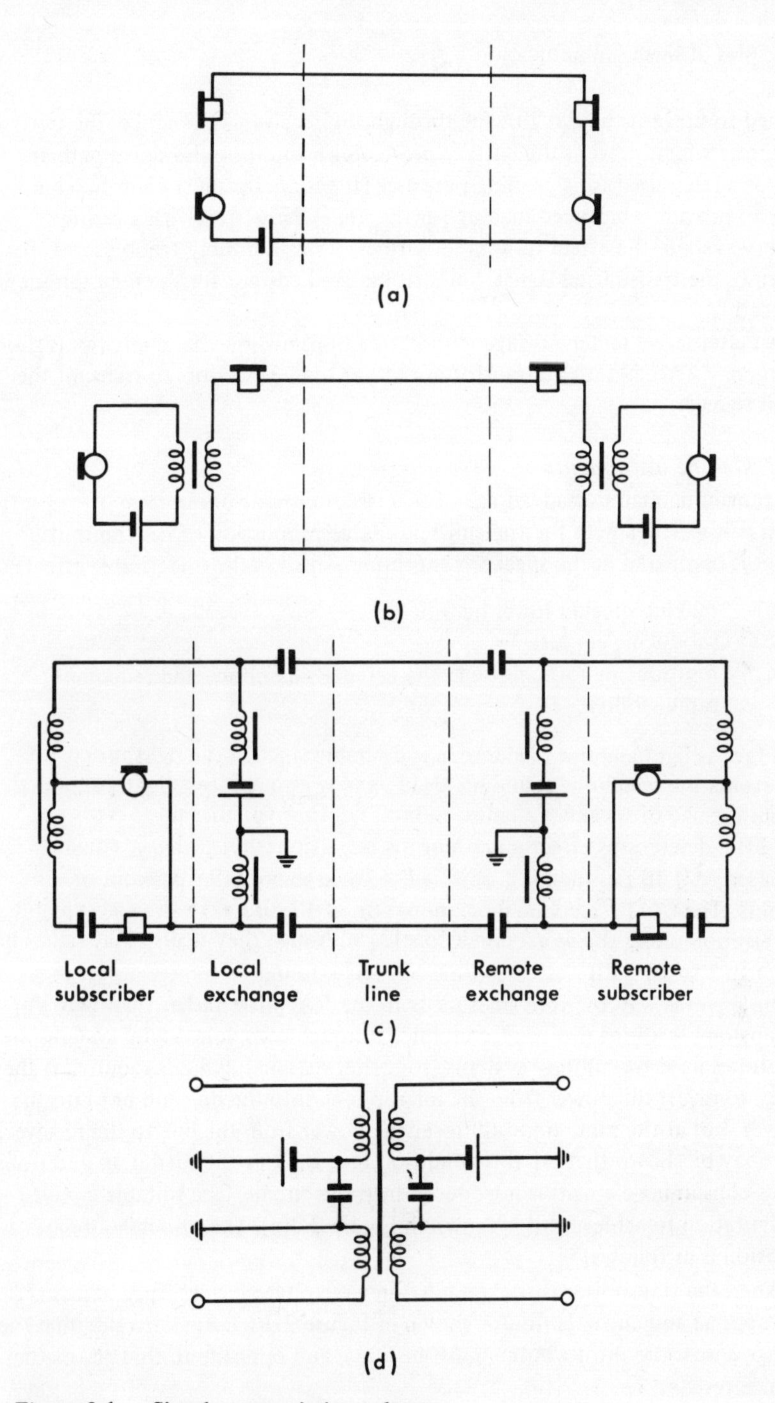

Figure 2.1 Simple transmission schemes
(a) Simplest form of two-wire speech link, (b) Use of transformer to increase power matching, (c) Use of central battery with capacitor feeding bridge (d) Transformer feeding bridge

needed to prevent the d.c. flowing through the receiver. In practice, the two inductors which prevent the battery producing a shunt on the signal path are part of a relay used for signalling purposes (the inductor must be in two halves so as to provide a balanced loading for the lines). Note that with a central battery system, the effect of long subscribers' lines is doubly troublesome, for not only the transmitted signal, but also the feed current for the microphone is reduced.

An alternative feeding bridge with better transmission characteristics is shown in Figure 2.1(d) and uses a transformer rather than capacitors to transmit the audio signals.

Anti-sidetone instruments

The remaining major disadvantage of this system is that of *sidetone*. Because the receiver is in series with the transmitter, a large proportion of the transmitted power is dissipated in the speaker's earphone. This has the undesirable effects of:

(a) making speaker lower his voice,
(b) producing ear fatigue,
(c) reproducing room noise in the listener's earphone and reducing intelligibility.

In fact a slight amount of sidetone is desirable since in the right proportion it prevents the telephone sounding 'dead', and it also reduces slightly the total dynamic range of speaker volumes. In fact, sidetone volumes up to 8 dB r.e. have little deleterious effect under normal conditions, but if the sidetone is increased to 0 dB r.e. then it is objected to by as many as 18 per cent of subjects. The CCITT recommend a minimum of 17 dB r.e. to allow acceptable conversation under the worst conditions [3] although they realise that values in the range 7 to 10·5 dB r.e. are more likely to be found.

The obvious way to avoid sidetone is to use four wires rather than two, but for most applications this is uneconomic, although complete 4-wire systems are sometimes used for military systems (for other reasons). What is required is the ability to divert the power from the microphone into the line and not into the receiver, but at the same time still receiving power from the line to the receiver.

It may be shown that [4] this condition may only be satisfied if an additional power-consuming element is introduced into the circuit. One suitable circuit configuration to achieve this is shown in Figure 2.2(a). The principle of operation is as follows [5].

When the transmitter is spoken into, an a.c. voltage is developed across the microphone and currents flow as shown in Figure 2.2(b). For zero sidetone the voltage across the third winding must be equal and opposite to that across the balance resistor, i.e.

$$rv = Bi_2$$

where v is the voltage across the first winding (this will be proportional to

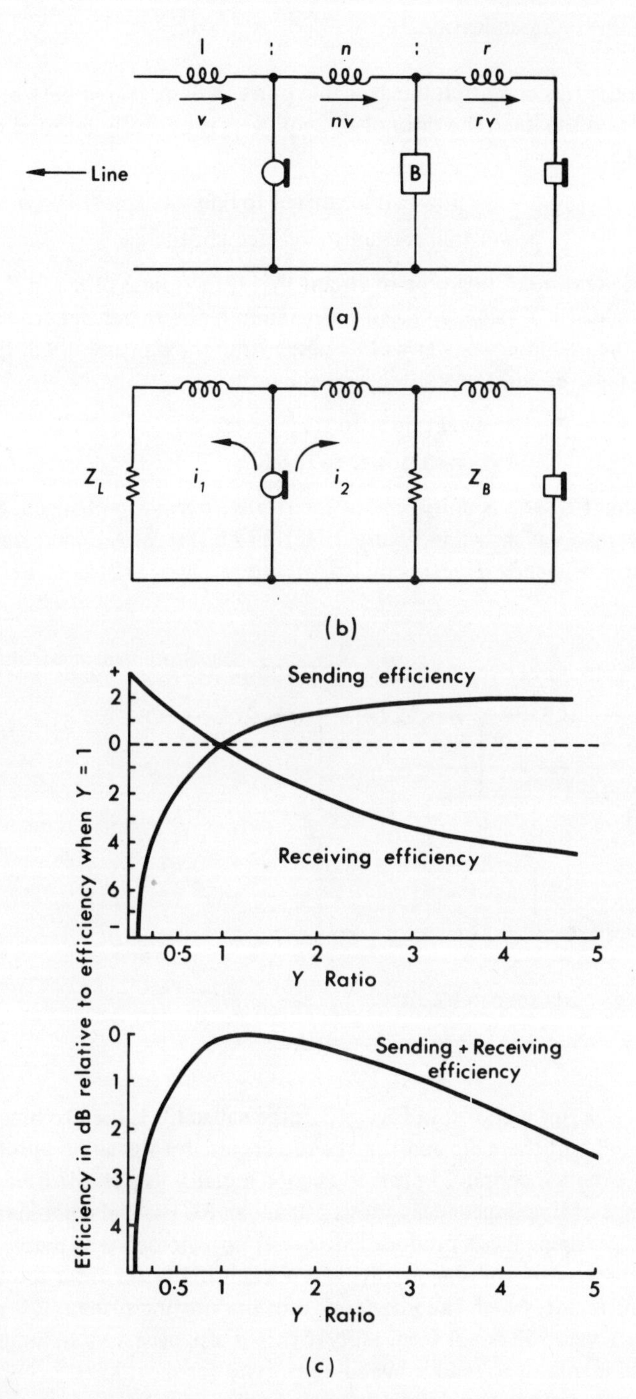

Figure 2.2 Basis of anti-sidetone circuits
(a) An anti-side tone circuit, (b) Transmitting condition, (c) Behaviour of a perfect telephone

$ni_2 - i_1$). Under this condition the available power will be shared between the balance load and the line. The ratio of these powers is normally referred to as the Y ratio, i.e.

$$Y = \frac{\text{power transmitted to line}}{\text{power transmitted to balance impedance}}$$

It is a basic feature of this type of circuit that if the values of n and Z_B are chosen for a specific Y value, and also for maximum power transfer from the transmitter, the maximum power will be taken from the line and it will split between the receiver and transmitter with the ratio

$$\frac{\text{Received power to transmitter}}{\text{Received power to receiver}} = Y$$

The optimum Y ratio is unity, and as the relative balance is changed, the total efficiency decreases as shown in Figure 2.2(c). For instance, if Y increases from 1 to 5, sending efficiency increases by 2·2 dB but receiving efficiency decreases by 4·8 dB.

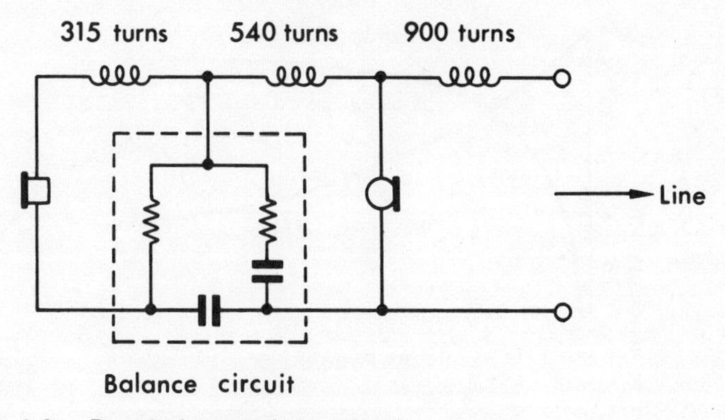

Figure 2.3 Practical transmission circuit

A practical circuit is shown in Figure 2.3; the capacitor is used to block d.c. In a practical circuit there are additional contacts etc. for signalling purposes. The balance circuit is chosen to provide as good a match as possible over the expected range of line impedances and is usually an RC parallel combination.

Modern development has produced a receiver of increased sensitivity, whereas the transmitter has improved in quality but not sensitivity. In order to utilise these for instruments which can interwork into the existing system, it is necessary to use a Y value different from unity [6]. It is also necessary to incorporate some form of regulator to reduce overall sensitivity for short lines.

It is also economic in some circumstances to have some amplification in the local handset, but the amount that it is practical is severely limited by cross-talk considerations [7].

2.2 Line theory

The lines connecting individual subscribers to the local exchange are the most expensive portion of the network, so any advance which can reduce their gauge is of great economic value. Typical figures for the British Post Office system are that the average investment per subscriber is £254, which is split up as follows:

Local network	£50	(20%)
Exchange equipment	£72	(28%)
Trunk and junction network	£59	(23%)
Subscribers' apparatus	£37	(15%)
Accommodation	£35	(14%)

i.e. nearly 20 per cent is spent in the local network. The gauge of the wire has traditionally been measured in lb/mile, but nowadays they are measured in mm diameter and the common ones in use in the United Kingdom are shown in Table 2.1.

Table 2.1

Conductor weight lb/mile	Diameter in mm	Resistance Ω/mile	Attenuation at 1·6 kHz in dB/mile
4	0·404	440	3·3
6½	0·508	270	2·7
10	0·635	176	2·2
20	0·902	88	1·5

At present the lines are used for individual subscribers, or a number on a party line. However, it is found that most underground lines can in fact transmit up to 150 kHz for distances of over a mile. The cross-talk increases with frequencies above 100 kHz but below this the spectrum may be used for simple f.d.m. systems [8] and other purposes, such as:

(a) Centralised electricity metering (20 kHz region).
(b) Civil defence messages and signals (72 kHz).
(c) Fire and intruder alarms (104 kHz).
(d) Recorded music distribution (144 kHz).

An increasing proportion of underground cables, particularly for the junction network, are being pressurised in order to improve their water-proofing properties. If a small leak does occur then the water is kept out by the escaping air which may also be detected at the exchange.

For the future, there is an interest in the use of aluminium conductors rather than copper, since the cost of aluminium is more stable [9]. The problems of jointing aluminium in the field have been solved, and in the United Kingdom aluminium cables are to be used increasingly for the local network.

An excellent summary of the practical problems of the planning and construction of local line networks may be found in reference [10].

Basic equations of line theory

The behaviour of telephone lines can be explained in terms of transmission line theory. If R, G, L and C are the resistance, conductance, inductance and capacitance per unit distance, then the propagation constant is given by [11]

$$\gamma = [(R + j\omega L)(G + j\omega C)]^{1/2} \tag{2.1}$$

and the characteristic impedance Z_0 given by

$$Z_0 = \left[\frac{R + j\omega L}{G + j\omega C}\right]^{1/2} \tag{2.2}$$

If $\gamma = \alpha + j\beta$ then the real part of the propagation constant, α, is given by

$$\alpha = [\tfrac{1}{2}\{|\gamma|^2 + (RG - \omega^2 LC)\}]^{1/2} \tag{2.3}$$

The behaviour of this gruesome expression may be discovered by investigating its form in several frequency regions.

(a) Low frequencies $\omega L \ll R$ and $\omega C \ll G$ Then

$$\alpha = (RG)^{1/2} \tag{2.4}$$

i.e. constant

(b) Intermediate frequencies $G/C \ll \omega \ll R/L$ (this is the physically reasonable case since L and G may usually be neglected for normal lines and hence $G/C \ll R/L$).

$$\alpha = (\tfrac{1}{2}\omega CR)^{1/2} \tag{2.5}$$

and also it may be shown to be greater than RG.

(c) Very high frequencies $\omega \gg R/L$, $\omega \gg G/C$,

$$\alpha = \tfrac{1}{2}\left[\frac{R}{Z_0} + GZ_0\right] \tag{2.6}$$

i.e. nominally constant but R and G vary with frequency due to skin effect. This will be discussed in Chapter 4.

It is clear that increasing R or G will increase the attenuation. However, if the expression for α in equation (2.3) is minimised with respect to L then a minimum value is found for

$$L = CR/G$$

or

$$\alpha = (RG)^{1/2}$$

which is the same as low frequency attenuation. Hence if L can be increased to have the value CR/G then the attenuation can be reduced, for all frequencies, to that of the low-frequency attenuation. This is called *loading* the line. With this condition the characteristic impedance becomes

$$Z_0 = \left(\frac{R}{G}\right)^{1/2} = \left(\frac{L}{C}\right)^{1/2}$$

which is a constant provided that R and G are independent of frequency. This implies that there is no phase distortion on the line and for this reason the condition is called the *distortionless line*. The velocity of propagation in this case is given by

$$v = (LC)^{-1/2}$$

so that as L is increased then this will be lower than for an unloaded line.

A similar result may be obtained if the expression for α is minimised with respect to C. This gives a minimum if

$$C = LG/R$$

but in general C is normally much greater than this value.

Thus if the inductance per unit length of the line can be increased then it is possible to reduce the attenuation at higher frequencies, giving a distortionless signal at the end, but with a slower velocity of propagation. Early attempts to achieve this condition were made by wrapping iron wire around the cables, but this proved difficult and expensive.

In practice the value of loading is too high to be provided economically but beneficial results are obtained even with values less than this. If the effects of G are negligible for the frequencies of interest then from (2.3) (i.e. $G^2 \ll \omega^2 C^2$)

$$\alpha = [\tfrac{1}{2}\omega^2 LC]^{1/2} \left[\left|1 + \frac{R^2}{\omega^2 L^2}\right|^{1/2} - 1\right]^{1/2}$$

so for $\omega L \gg R$,

$$\alpha = \tfrac{1}{2}R\left[\frac{C}{L}\right]^{1/2}$$

Hence α may be made smaller by increasing L, and provided $\omega L > \tfrac{1}{2}R$ then this value of α is smaller than for the case of negligible L and G, which is the value in equation (2.5).

At the turn of the century Pupin and Campbell almost simultaneously discovered that the same effect could be achieved by adding lumped inductance at regular intervals. This has exactly the same effect as the continuous loading up to a critical cut-off frequency. Beyond this frequency the attenuation increased sharply, Figure 2.4. (In fact it was this discovery which led to the invention of filters). This cut-off frequency was dependent upon the spacing. For instance, 88 mH coils at 1·6 km spacing will give a cut-off frequency of about 3·5 kHz on 20 lb/mile twisted pair. 22 mH coils every 0·4 km would give a cut-off of around 14 kHz. However, the cost in propagation delay is high, around 22 000 km/s compared to 220 000 km/s for a cable carrier system and 300 000 km/s for radio links. Thus even for 160 km of loaded cable there will be a one-way propagation time of about 7 ms.

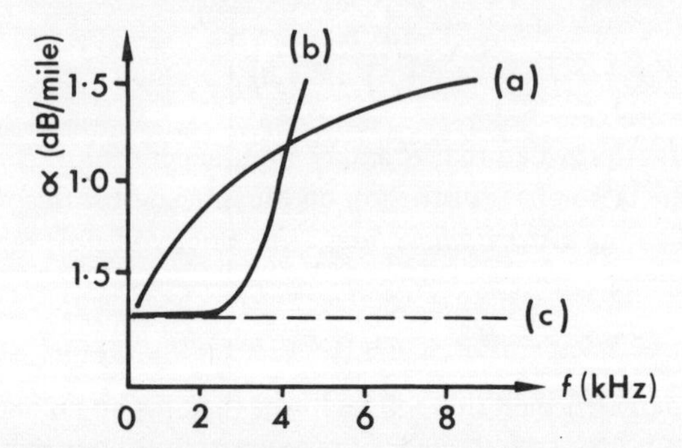

Figure 2.4 Effect of loading coils
(a) basic line, (b) lumped loading, (c) continuous loading

The actual form of the inductors is a toroid wound as shown in Figure 2.5. The inductors must be stable with time and temperature, have a low residual field and pick-up, and be small and cheap. They must also have a low d.c. resistance. This type of loading is used extensively on short distance junctions, 70 per cent of which are presently loaded although this percentage will decrease once p.c.m. systems are introduced. They are also used sometimes on the local distribution side for long subscribers lines [12].

Figure 2.5 Winding of loading coil

Phantom line working

It is possible to increase the number of channels using a fixed number of wires if the technique of *phantoming* is used. This technique is illustrated by Figure 2.6(a). If two twisted pairs are themselves twisted together then they may form a third twisted pair (this is called a *multiple* twin). Any voltage that is impressed upon them will be the same for both wires and hence there will be no net effect. This assumes that the cables are well balanced, otherwise cross-talk will occur. If phantoms are used, they have to be loaded separately as shown in Figure 2.6(b).

The most common use of phantoms is on carrier cables (see Chapter 4) where the phantom is used to carry programme channels. In practice the usual arrangement is to twist all four wires together in one bunch with the opposite wires forming a pair. This is called a *star-quad* and produces a better balance and takes up less room in the cable than a multiple twin. The twist lengths of adjacent quads are arranged to be different in order to reduce the cross-coupling.

Double phantoms are also used but for poor quality circuits such as engineers' test lines, signalling or power supplies etc.

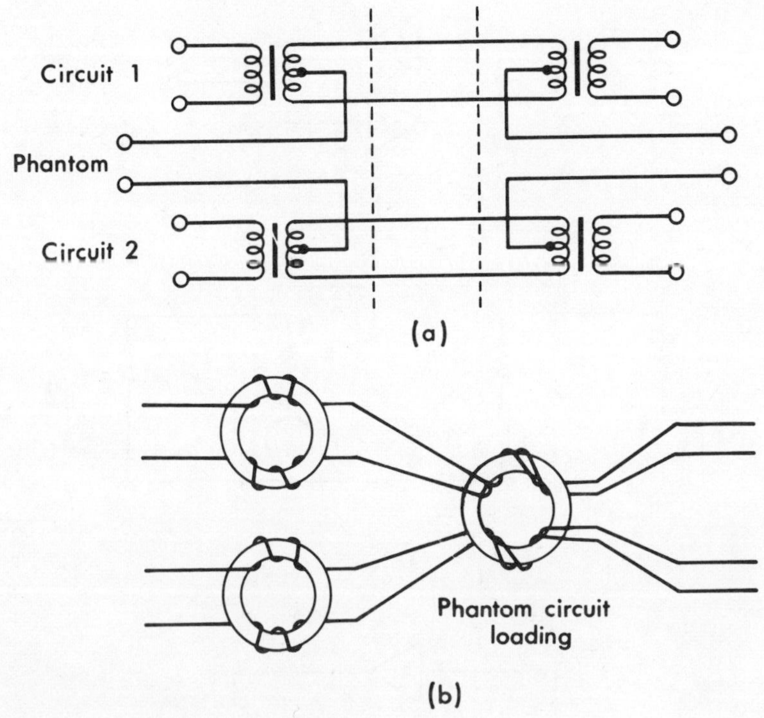

Figure 2.6 Phantom circuits
(a) Basic principle, (b) Loading arrangements

2.3 Audio amplification

Hybrid transformer–2- to 4-wire convertors
In a two-wire system, the signals are passing in both directions at once and in the same frequency band. Hence a bi-directional amplifier is necessary if any amplification is required. The simple system shown in Figure 2.7(a) will not work because although it can be made stable, it will not produce any insertion gain, as a detailed analysis of the equivalent circuit of Figure 2.7(b) would show.

What is needed is a system whereby the output from A_1 is sent to the line but prevented from reaching the input of A_2 and similarly for the output of A_2 as shown in Figure 2.7(c). This might seem at first sight to be impossible, but it may be achieved by use of the same principle as that employed for the anti-side-tone circuit. One form used is that of the *hybrid transformer* shown in Figure 2.8(a). This consists of two cross-coupled transformers together with a network, N, whose impedance is the same as that of the line, Z, over the frequency band of interest. The main requirement of this circuit is that none of the output from A_2 should reach the input of A_1. This output will generate equal voltages across windings W_1 and W_2, and provided $N = Z$, equal currents will flow in the

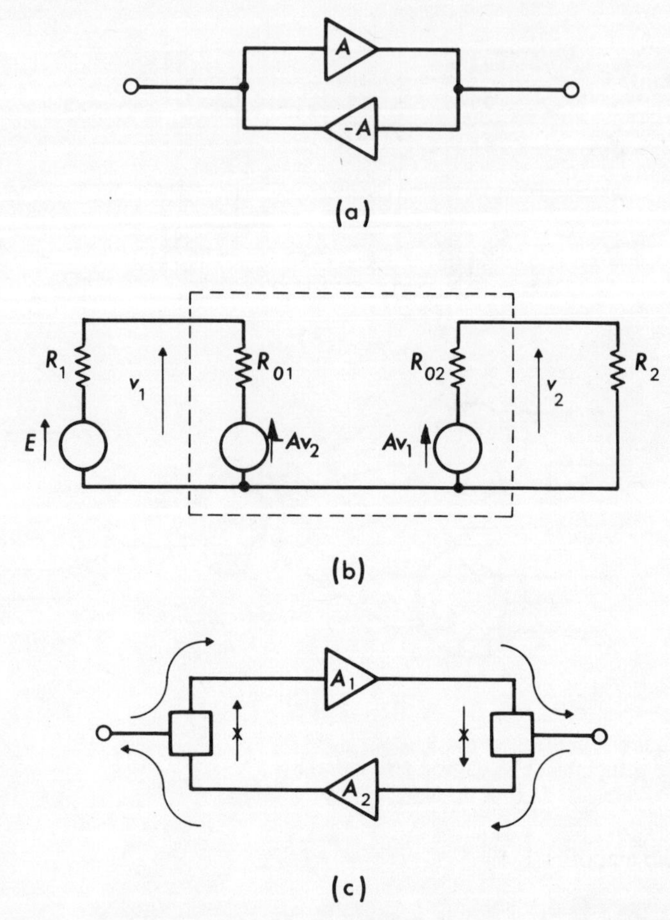

Figure 2.7 Problems of bi-directional amplification
(a) simple, but ineffective bi-directional amplifier (b) Equivalent circuit used for computing insertion gain (c) Basic requirement for stability and gain

line and balance circuits. Since one of the windings, W_5, is connected in anti-phase to W_4, there will ideally be no net m.m.f. on T_2 and hence no signal at the input of A_1. Hence, the power from A_2 will divide between N and Z. This division will be equal since N and Z are equal.

Power from the line will divide between the input of A_1 and the output impedance of A_2 (where it will have no effect); none will go to N. The price paid for the separation is then a 3 dB loss in each direction. Although the loss in one direction could be reduced, it will be at the (disproportionate) increase of loss in the other direction. Also, since transformers are used, the d.c. signalling path is removed. This then achieves the required behaviour but it is critically dependent upon the equality of N to Z, and since the range of impedances that a switched

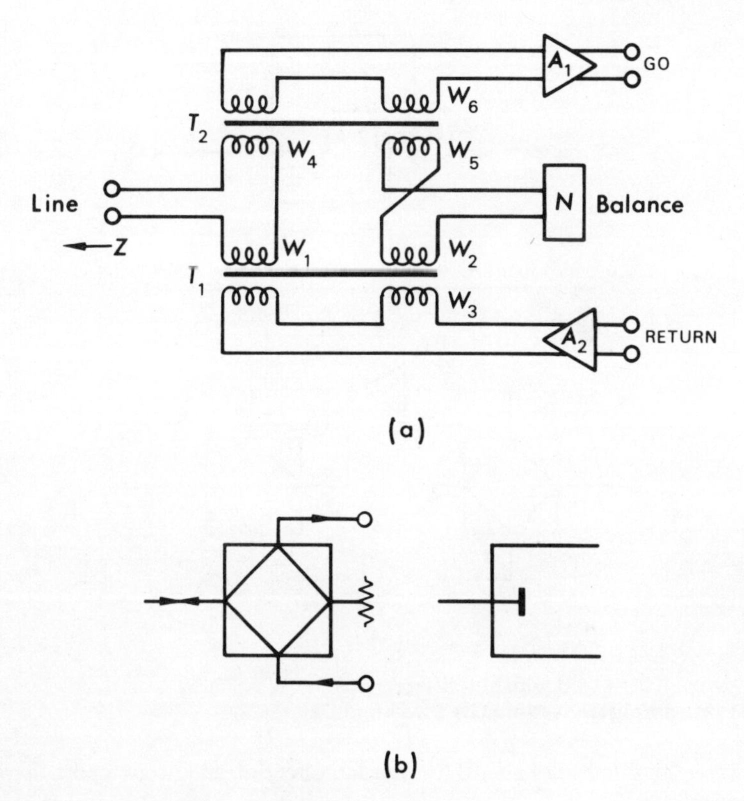

Figure 2.8 2–4 Wire transformer
(a) Hybrid transformer, (b) Symbols for a hybrid

line may present to this network is large, there will always be some power transfer across the transformer, the implications of which are considerable and will be discussed later.

The usual symbol for this hybrid is shown in Figure 2.8(b) and for obvious reasons it is frequently referred to as a 2-4 wire convertor. Another form of a hybrid is shown in Figure 2.9 and this uses resistors rather than a transformer.

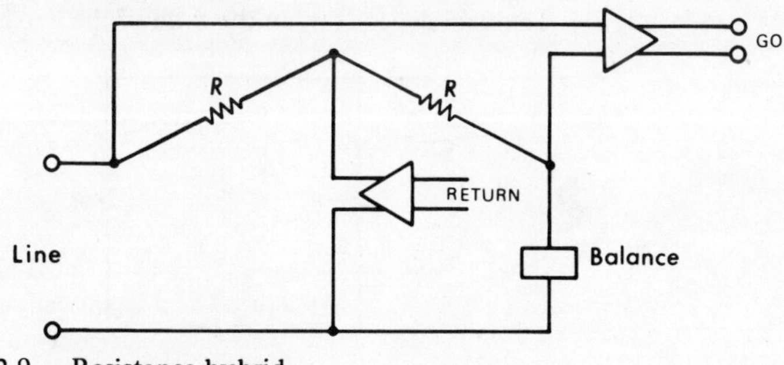

Figure 2.9 Resistance hybrid

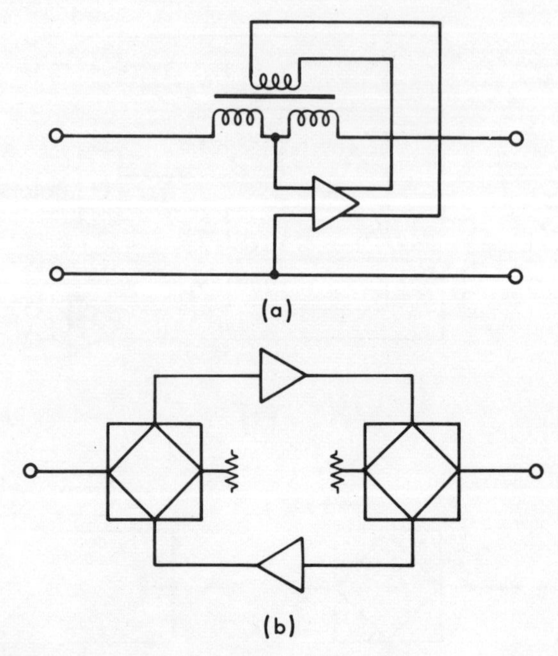

Figure 2.10 Two-wire amplifiers
(a) One-amplifier version, (b) Two-amplifier version

However, this circuit has a 6 dB loss in each direction and no two ports have a common point.

If a two-wire amplifier is needed then it can be provided by a single amplifier and a single transformer as shown in Figure 2.10(a), but in practice better stability may be obtained by the use of two amplifiers and hybrids as shown in Figure 2.10(b).

Negative impedance convertor

An alternative solution is a two-terminal bi-directional amplifier which is based on the concept of negative impedance. It may be shown that if a negative resistance is connected across a line as, shown in Figure 2.11, then power gain may be produced [13]. This gain depends upon the ratio of the line impedances

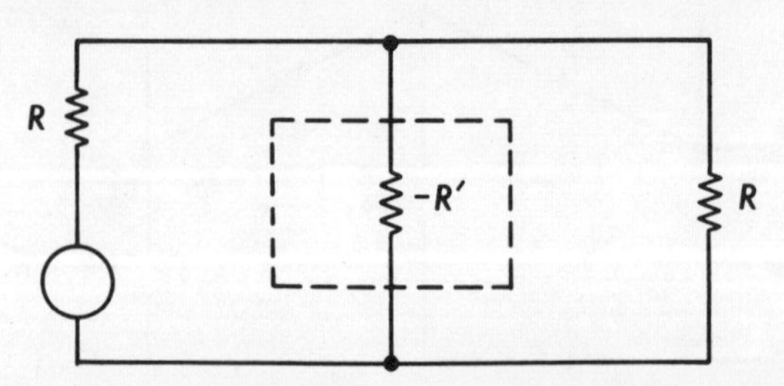

Figure 2.11 Use of negative impedance for amplification

and the negative impedance produced and its stability depends upon these parameters as well. Its main use is for unloaded cicuits up to 40-50 km length, but due to the impedance matching problem, the gain is limited to about 10 dB, and generally not more than one can be used (near the centre of the route). Compared to the hybrid repeater, its main advantages are:

(a) cheaper,
(b) smaller,
(c) d.c. path maintained,
(d) less current required,
(e) signal path still maintained if circuit fails.

Minimum realisable losses of 2-wire amplified circuits
The effect of finite power transfer across a hybrid transformer due to the impedance mismatch of the line and balance network will be shown to have the effect of placing a lower limit on the minimum loss at which the complete circuit may operate. The trans-hybrid loss may be conveniently considered to consist of two parts, as shown in Figure 2.12(a).

(a) The loss due to the nature of the hybrid transformer is 3 dB between the RETURN amplifier and the 2-wire point, plus another 3 dB from the 2-wire point to the GO amplifier. In practice the 3 dB loss through the hybrid is nearer 3·5 dB due to transformer inefficiencies.

(b) The *balance return loss* (B_S dB) which is defined as that portion of the total transmission loss introduced by the hybrid transformer between GO and RETURN amplifiers which is attributable to the degree of impedance match between the line and the balance network. It is given approximately by

$$B_S = 20 \log_{10} \left| \frac{Z + N}{Z - N} \right| \text{ dB}$$

where Z is the line impedance and N is the impedance of the balancing network. The expression is exact if the balance network, the input impedance of the GO amplifier and the output impedance of the RETURN amplifier are all equal. The derivation of the exact expression together with a physical interpretation of the concept is contained in Appendix A.

The total trans-hybrid loss is thus (B_S + 6) dB for an ideal transformer and (B_S + 7) dB in practice. Note that for $Z = N$, B_S is infinite and if Z is an open- or a short-circuit, then $B_S = 0$ dB.

Consider now a complete amplified circuit as shown in Figure 2.12(b). Assume the loss between the 2-wire points X and Y is T dB; this includes the losses caused by the hybrid plus the effect of the amplifier gains and any line losses. The total loss around the loop will be $M = 2(T + B_S)$ dB. If at any

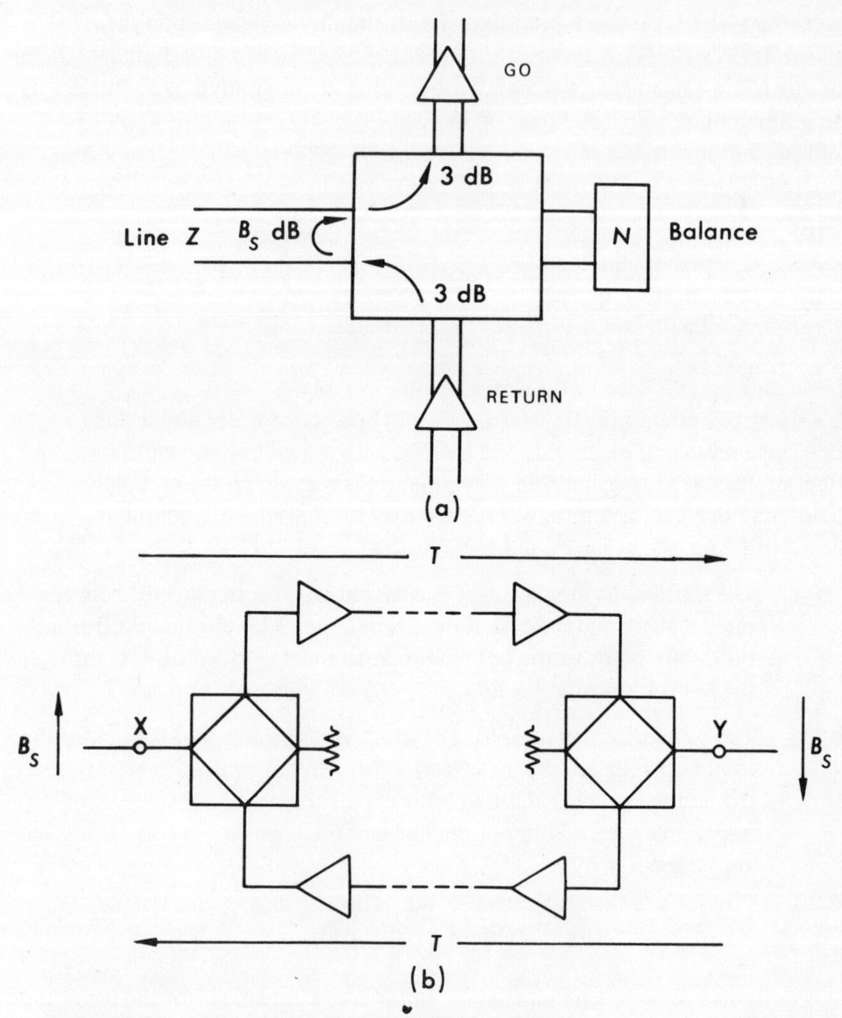

Figure 2.12 Use of balance return loss concept
(a) Definition of trans-hybrid loss, total loss between RETURN and GO is (B_S + 6) Db; (b) Amplified 4-wire circuit, loop loss is $2(T + B_S)$ dB

frequency $M = 0$ dB (i.e. unity gain) then oscillation will occur (or 'sing' as it is called). Even if $M > 0$ dB, the circuit may ring and this will cause undesirable distortion to the signal. In order to avoid this condition it is found that the total loop loss should be greater than 6 dB (this is frequently called the *singing margin*). Hence, for a given minimum value of balance return loss, there is a lower limit to the transmission loss, i.e.

$$2(T + B_S) \geqslant 6 \text{ dB}$$

and so

$$T \geqslant 3 - B_S \text{ dB}$$

In the limiting case if the range of line condition presented to the hybrid transformer can range from open to short circuit and anything in between, then the minimum value of B_S will be 0 dB and hence the minimum loss at which the circuit may be operated is 3 dB.

The balance return loss at the hybrid transformer may be improved by adding a matched attenuator on the 2-wire side as shown in Figure 2.13 which reduces the effect of the impedance variations on the line. If the attenuation produced is A dB then it may be shown that the improvement in the balance return loss is at most $2A$ dB. For the case shown in Figure 2.13 this means that T may be reduced by $2A$ dB for the same stability margin. However, the overall loss between the new 2-wire points, T', will have increased by $2A$ dB and hence there is no overall reduction in the minimum loss.

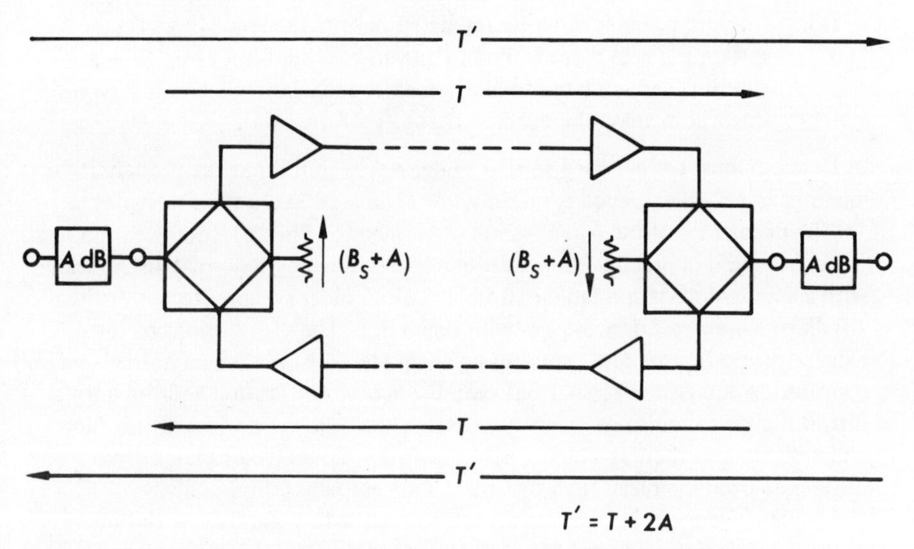

Figure 2.13 Effect of adding attenuation to lines to improve balance return losses

If the attenuation is in fact produced by a length of cable which is well matched to the hybrid transformer, then the same result will be produced. Thus if one can control the image impedances of trunk and junction cables then it is possible to reduce their overall loss by the use of hybrid transformers and amplifiers situated at intervals along the cable. However, it is not possible to decrease the minimum loss between the points at which one must expect large impedance variations, i.e. at the switching points.

This result is in fact only a particular example of a fundamental property of systems exhibiting bi-directional gain [14]. Stated simply, this property is that if a two-port system must be stable under all combinations of termination then the product of its gains in both directions must be less than unity, i.e. if the loss in one direction is T_1 dB and in the other is T_2 dB then for absolute stability

$$T_1 + T_2 > 0 \text{ dB}$$

The derivation and discussion of this property is contained in Appendix B.

The effect of statistical variation in T

An additional factor in the calculation is the accuracy to which T can be specified. Its nominal value can vary due to:

(a) Variation in equipment gain with time and temperature.
(b) Inaccuracy of alignment on part of maintenance staff.
(c) Inaccuracy of test gear.
(d) Variation of gain of circuits between measurement frequency (usually 800 Hz) and other speech frequencies.
(e) Number of independently regulated links in tandem. (In an f.d.m. system the gain is controlled by means of a pilot but every time a circuit is demodulated to audio then the regulation is terminated and this will increase the overall variability of the gain).

Hence T must be increased so that under worse conditions the probability of the singing margin exceeded is suitably low (1 in a 1000 is a typical figure).

The departures of the actual values of T (measured in dB) from their nominal value are found in practice to be adequately described by a normal distribution with a standard deviation ranging from 1·5 dB in older systems, to better than 1·0 dB in a modern f.d.m. system with regulation. The distributions of the balance return loss are also found to be adequately described by a normal distribution and any difference between the actual distribution and the normal distribution is found to be on the pessimistic side (i.e. the normal distribution will indicate that there is a higher proportion of values greater than a given difference from the mean than the proportion actually found).

Hence if the mean balance return loss is B dB with a standard deviation of σ_B dB and the mean value of loss is T dB with a standard deviation of σ_T on each independent link, then the mean stability margin is

$$S = 2B + 2T \text{ dB}$$

with a standard deviation of

$$\sigma_S = [2\sigma_B{}^2 + n(2\sigma_T)^2]^{1/2} \text{ dB}$$

where it is assumed that the values of balance return loss at each end are uncorrelated, that there are n independent links, and in any link the losses of the two directions are correlated. This last assumption is the most pessimistic.

Since the distribution of the stability margin is normal then the value that will be exceeded 99·9 per cent of the time is $S - 3{\cdot}09\sigma_S$, so for a 0·1 per cent probability that the stability margin is below the singing margin then

$$S - 3.09\sigma_s \geqslant 6 \text{ dB}$$

As a practical example, assume that the mean value of B is 6 dB with σ_B = 2.5 dB and that σ_T = 1 dB. Hence

$$12 + 2T - 3{\cdot}09[12{\cdot}5 + 4n]^{1/2} > 6$$

i.e.

$$T > 1{\cdot}54[12{\cdot}5 + 4n]^{1/2} - 3 \text{ dB}$$

The values of T are shown in Table 2.2.

Table 2.2 Minimum Stability Values

n	1	2	3	4	5	6	7	8
T_{min}	3·3	4·0	4·6	5·2	5·8	6·3	6·8	7·3
$4 + 0{\cdot}5n$	4·5	5·0	5·5	6·0	6·5	7·0	7·5	8·0

In practice a simple rule would be used in cases like this; the usual rule is

$$T = 4{\cdot}0 + 0{\cdot}5n \text{ dB}$$

and as may be seen from Table 2.2 this gives a more than adequate margin of stability.

Concept of the zero reference point and relative level

In any system involving amplifiers and losses it is convenient to define some point (usually the input) as a *zero reference point*, and measure the relative gains and losses at different points of the system relative to this point. The gains or losses of the system are normally expressed in dB and the unit of relative level is called the dBr, i.e. if at some point in the system a signal is 10 dB higher than it is at the zero reference point then the relative level at this point is said to be 10 dBr. An example is shown in Figure 2.14(a).

Conversely, if the relative levels of a system are known, and so is the signal power at the zero reference point, then the signal power at any other point may be found. For instance, if the signal power at the zero reference point is 5 dBm, then at a point of relative level of 10 dBr, the signal power will be 15 dBm. The concept of relative level is also very useful for referring observed powers back to the input; if a signal power of 10 dBm is observed at a point whose relative level is –8 dBr, then this would correspond to a power at the zero reference point of 18 dBm. When a power is referred back to the zero reference point in this manner the units are expressed in dBm0. This may be summarised:

'dBm' = 'dBm0' + 'dBr'

When these concepts are applied to a 4-wire amplified circuit, the zero reference point is usually taken to be the 2-wire input to the hybrid transformer. Since the two directions of transmission are essentially independent, then each direction will have its own zero reference point. An example is shown in Figure 2.14(b) where it is assumed that the hybrids introduce a 3·5 dB loss. Note that it is not necessary for the two paths of the 4-wire system to have equal gains.

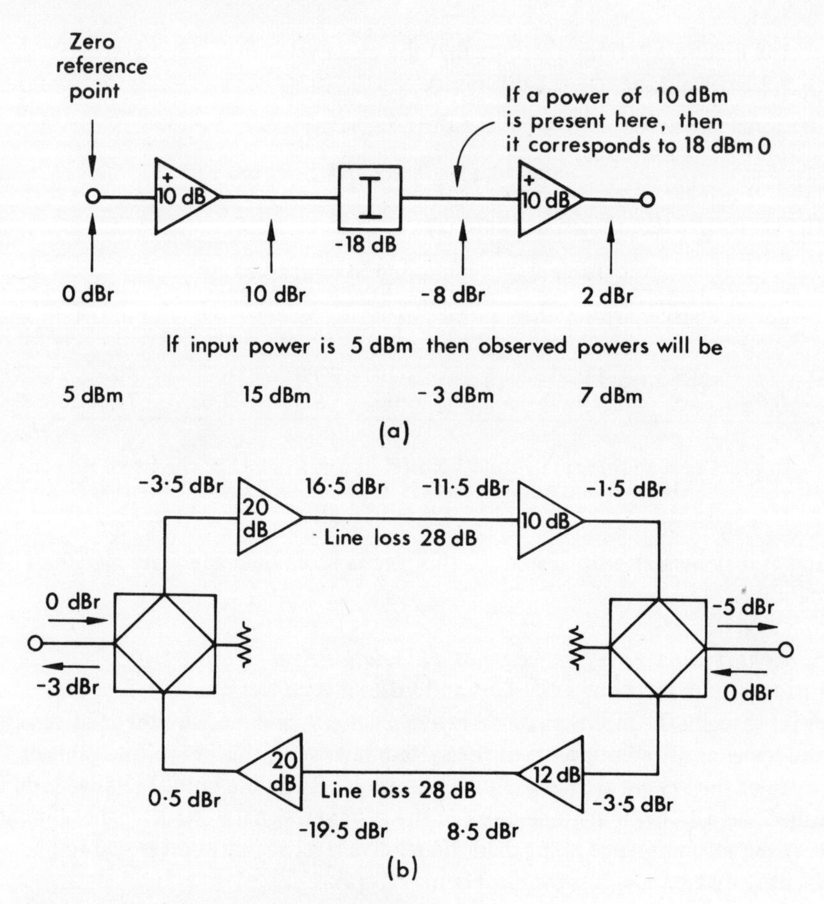

Figure 2.14 Concept of relative level
(a) 2-wire circuit, (b) 4-wire circuit

If a system is 4-wire switched then the individual links between switching points will have no hybrid transformer which can act as a zero reference point. In these cases the relative levels are defined with respect to the input of a hypothetical hybrid transformer that would be present if the system were terminated. Thus each link in a switched system has its own point of zero relative level.

In North America the equivalent term to dBr is *transmission level point*; the reference point used in that country is at the output of the toll exchange. As a result of a gradual up-grading of transmission performance, trunks have had their loss decreased to 0 dB in some cases, and the stability obtained by the use of 2 dB pads in the exchange. Hence the output from the toll exchange is now –2 dB TLP.

In international use the relative levels are defined at a measurement frequency of 800 Hz, although 1000 Hz is still widely used.

2.4 The production and control of echos

Whenever there is an impedance mismatch in a transmission system, then there will be some energy reflected, and this energy may return to the speaker or reach the listener some time after the original signal has been sent or received. This will give rise to speaker or listener echo if the delay is greater than about 10 ms. In practice the most troublesome form of echo is that produced at the hybrid transformers as shown in Figure 2.15(a). As may be seen, the speaker echo is attenuated by $(B + 2T)$ dB at the hybrid 2-wire nearest the listener, and it will return to the speaker with a delay of $2(t_1 + t_2)$ seconds where t_1 seconds is the

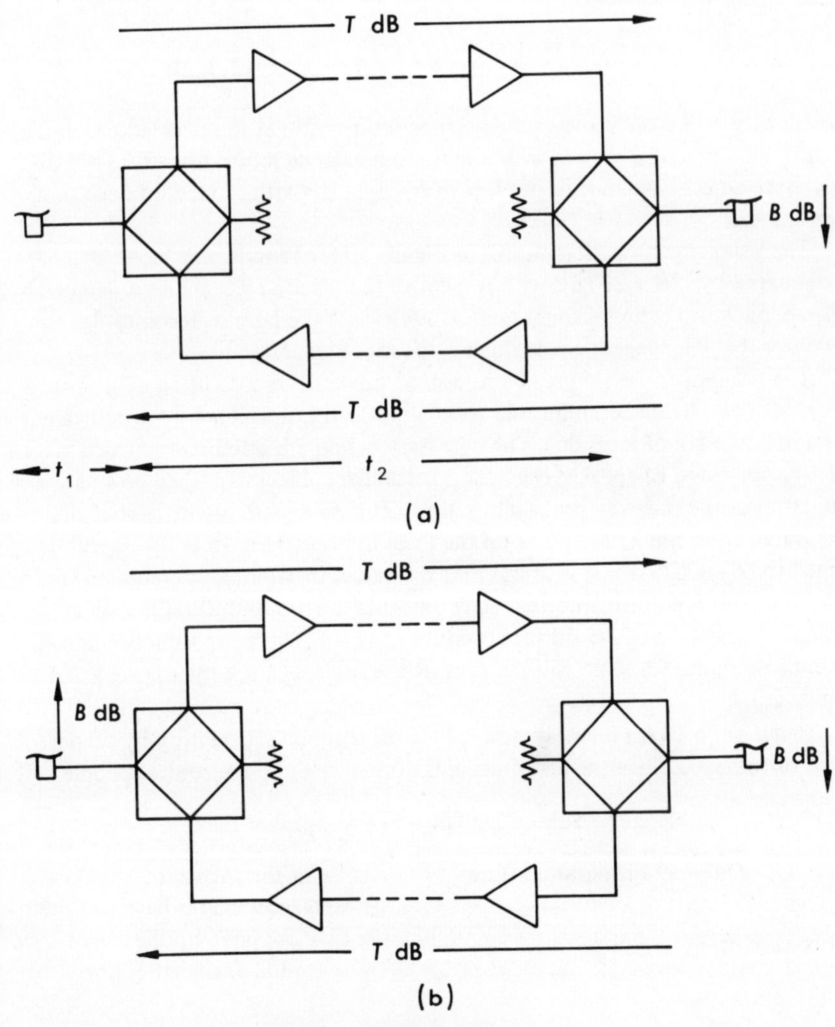

Figure 2.15 Echo paths in amplified circuits
(a) Talker echo path, (b) Listener echo path

delay from the handset to hybrid transformer and t_2 seconds is the delay between the 2-wire points of the two hybrids.

B is the balance return loss at the remote hybrid. The value used for B is not the same as that used for stability calculations, since in this use it is the minimum value of B in the audio range that governs the stability, whereas from the echo view-point it is the average value of B in the audio-band that is important. The CCITT have provisionally recommended[15] that the value of B for the echo point of view be taken as the unweighted mean of the power in the band 500-2500 Hz. A convenient method of calculating this mean is to divide the band by five equally spaced ordinants and apply Simpson's rule for numerical integration, i.e.

$$B_E = 10 \log_{10} \frac{1}{4}\{\tfrac{1}{2}b_1 + b_2 + b_3 + b_4 + \tfrac{1}{2}b_5\} \text{ dB}$$

where b_1 etc. are the value of the balance return ratio at the selected ordinants. Typically a practical system with a minimum balance return loss of $B_S = 3$ dB will have an echo balance return loss of $B_E = 11$ dB with a standard deviation of up to 3 dB.

The listener echo path is shown in Figure 2.15(b) and it may be seen to be attenuated by $(2B + 2T)$ dB with respect to the received signal and is delayed by $2t_2$ seconds. An echo becomes more troublesome the later it becomes. In practice it is the speaker echo which is the more troublesome.

The tolerance to echo may be measured subjectively by setting up a connection with two handsets connected to the hybrid transformers by the equivalent of a short length of local line. The necessary echo-path attenuation which gives a satisfactory level of speaker echo for a particular subject may then be found as a function of the one-way propagation time. This echo-path attenuation is that measured from the 2-wire point on the local hybrid transformer, i.e. it will be equal to $(B_E + 2T)$ dB. It is found that the values of attenuation required for different subjects form approximately a normal distribution (in dB) with a standard deviation of 2·5 dB for all values of delay. The mean values obtained (i.e. those values for which 50 per cent of the subjects are satisfied) are given in Table 2.3.

Hence for a 10 ms one-way delay, 11·1 dB is found satisfactory by 50 per cent of subjects. Since the distribution is normal then 66 per cent of people will

Table 2.3 Subscriber Tolerance to Speaker Echo

One-way propagation time (ms)	Echo-path loss for average subject (dB)
10	11·1
20	17·7
30	22·7
40	27·2
50	30·9

be satisfied by a loss of the average plus one standard deviation, i.e. 13·6 dB. 99 per cent of the subjects are satisfied by a loss equal to the mean plus 2·33 times the standard deviation, i.e. 11·1 + 5·8 = 16·9 dB. For a small country such as the United Kingdom very few connections are going to have delays exceeding 10 ms. Hence if the value of the echo balance return loss is 11 dB then 2·9 dB would produce satisfactory echo performance for 99 per cent of the population.

However, this does not bring into account the statistical variation of B_E or of T. The mean margin against objectionable echo is given by

$$M = B_E + 2T - E \text{ dB}$$

where E is the mean threshold attenuation as given in Table 2.3. The standard deviation of the margin is

$$\sigma_M = [\sigma_B{}^2 + n(2\sigma_T)^2 + \sigma_E{}^2]^{1/2} \text{ dB}$$

assuming that there are n independent links and that the gain variations in the two directions of each link are correlated (the most pessimistic assumption). Hence for $\sigma_B = 2.5$ dB, $\sigma_T = 1$ dB and $\sigma_E = 3$ dB,

$$\sigma_M = [15{\cdot}3 + 4n]^{1/2} \text{ dB}$$

If it is desired to keep the margin positive for 99 per cent of the time then

$$M - 2{\cdot}33\sigma_M \geqslant 0 \text{ dB}$$

So, for example, a 10 ms one-way delay and nine links in tandem then the mean value of $B_E + 2T$ must exceed

$$B_E + 2T \geqslant 11.1 + 2.33 \times 7.2$$
$$\geqslant 27{\cdot}9 \text{ dB}$$

for $B_E = 11$ dB, $T \geqslant 8{\cdot}5$ dB and this is likely to be higher than the value used for stability. However, a 10 ms delay corresponds to about a 1200 mile connection (2000 km) and is unlikely to occur in a small country such as the United Kingdom. If it did then the transmission loss is very likely to exceed the 8·5 dB.

Echo suppressors

For longer delays, as are experienced on international connections and in North America, Australia, etc., the value of T required for echo control is greater than that required for stability and it is possible to increase the value of T to reduce the echo problem[16]. This technique is used extensively in the countries named for internal calls (see next chapter) but it is not applicable to long-distance international working since in these cases one cannot afford the extra attenuation, and echo suppressors are needed.

There are two types of echo suppressors in current use; one is for use in terrestrial systems where one-way propagation up to 50 ms is encountered, and a more sophisticated type[17], for use in very long inter-continental connections and calls using a synchronous satellite (270 ms one-way propagation time). The usual method of operation of an echo suppressor is to locate it as near as possible (in terms of delay) to the source of the echo, and it is then operated by signals from the remote end. When they detect a speech burst coming from the distant end, they are designed to insert a high loss into the return path in order to suppress the echo produced. There is one such echo suppressor (called a half-echo suppressor) at each end of the connection. The design is complicated by the economic need to locate the suppressors only where they are needed, and this usually implies the international exchange. However, the source of the echo may be in the national network, and hence it may be some time after the original speech burst has passed the suppressor before its echo returns. It is necessary to allow the listener to interrupt the talker at any point, and hence some design compromises are necessary.

The problems are even more acute for the long-delay suppressors and this has necessitated radical redesign of the suppressors. A block diagram of a modern suppressor is shown in Figure 2.16.

Other practical problems are:

(a) If the connection is used for data transmission then the echo suppressors are an embarrassment and arrangements must be made for a subscriber to disable any suppressors if he wishes to send data (see Chapter 12).

(b) In a long distance switched connection it might be possible to have several links in tandem, each of which could be provided with a pair of echo suppressors, and if these are all allowed to operate then various

Figure 2.16 Block diagram of modern echo suppressor (from A. G. Hodsall, 'Taking the Noise out of Satellite Calls', *Post Office Telecommunications Journal* 20, 2, pp.16–19, 1968)
(a) When B speaks to A, the echoes of his speech are suppressed by the 60 dB attenuation, switched into the return transmission path by the suppressor at A's end of the circuit. A small part of the speech signals from B pass via the bandpass filter (BPF) to be detected by the logic circuit which operates the 60 dB switch. Similarly, the suppressor at B's end prevents echoes being returned to A when he speaks. (b) A simplified block schematic diagram of the echo-suppressor at A's end. When B speaks, the suppression logic passes a signal through the logic circuit to operate the 60 dB suppression. The break-in speech detector continuously compares speech signals in two directions of transmission. When it detects break-in speech from A it passes a signal to the logic circuit which removes the 60 dB suppression, thus permitting A's speech to be transmitted to B. Under break-in conditions, the logic circuit also switches an attenuation of 6 dB into the receive transmission path. Because the suppressor at B has switched to the same state, there is a total attenuation of 12 dB switched into the echo paths which partially suppresses echoes when both A and B speak at the same time. A conditioning tone sent in advance of data signals is detected by the tone disabler which sends a signal to the logic circuit to disable operation of both the 60 dB and the 6 dB switches.

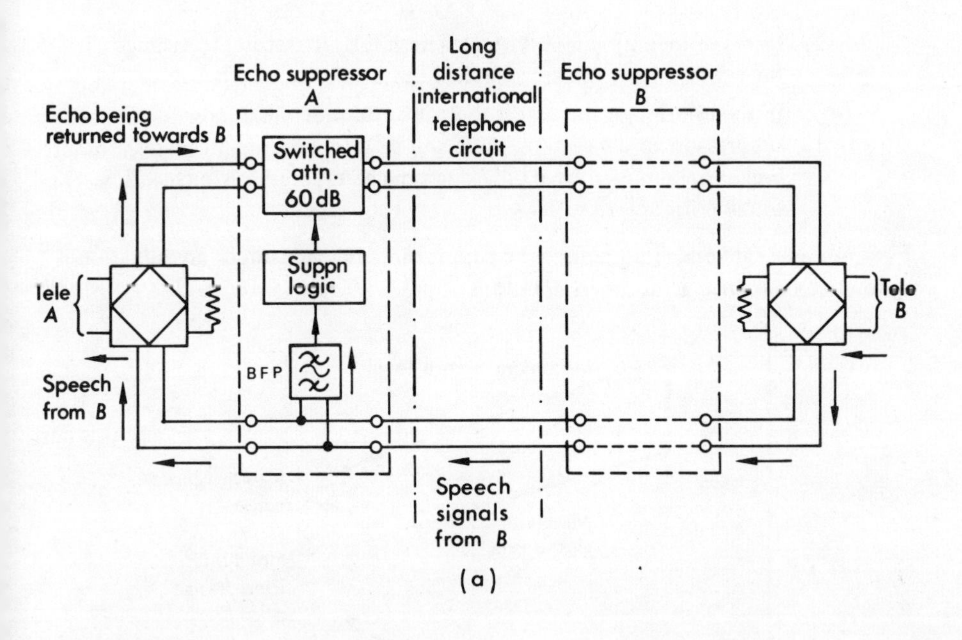
Echo suppressor
A
Long
distance
international
telephone
circuit
Echo suppressor
B
Echo being
returned towards B
Switched
attn.
60 dB
Suppn
logic
Tele
A
BFP
Speech
from B
Tele
B
Speech
signals
from B

(a)

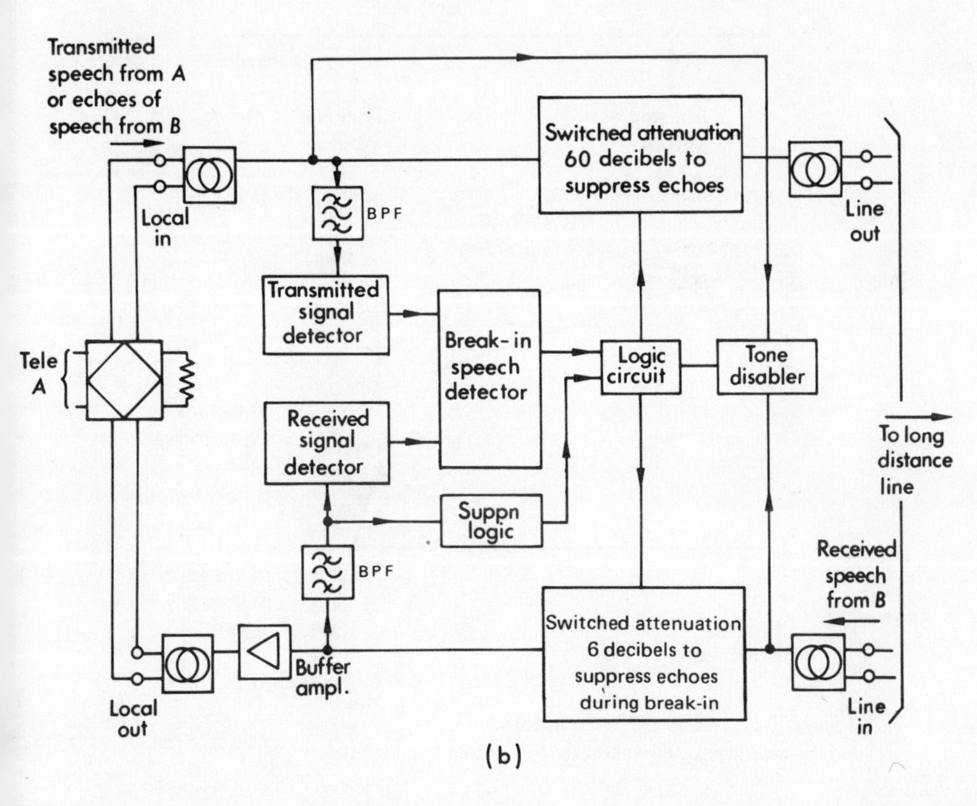
Transmitted
speech from A
or echoes of
speech from B
Local
in
BPF
Switched attenuation
60 decibels to
suppress echoes
Line
out
Transmitted
signal
detector
Break-in
speech
detector
Logic
circuit
Tone
disabler
Tele
A
Received
signal
detector
To long
distance
line
Suppn
logic
BPF
Received
speech
from B
Switched attenuation
6 decibels to
suppress echoes
during break-in
Buffer
ampl.
Local
out
Line
in

(b)

forms of lock-up may occur. It is desirable, therefore, to arrange that the echo suppressors are switched out at a transit exchange.

(c) In the interim period whilst there are still short- and long-delay suppressors in the system, there will be interworking problems. The long-term aim is to have only one type of suppressor which will be suitable for any time delay.

An alternative technique for echo suppression which is under investigation is to use an adaptive echo canceller[18].

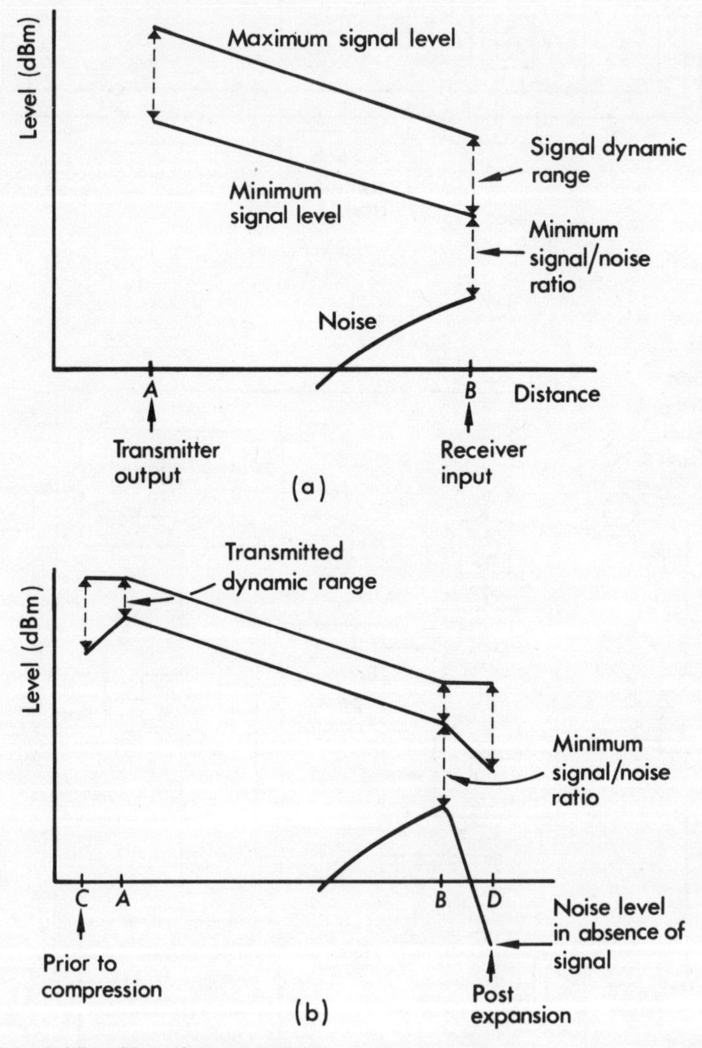

Figure 2.17 Use of compandors
(a) Non-companded system, (b) Companded system

2.5 Use of compandors to reduce effects of circuit noise

One technique which is of use for reducing the effect of circuit noise is that of companding. If, prior to transmission, a signal is passed through a non-linear circuit which has the effect of reducing the dynamic range of the signal (compressor), and the output level is adjusted so that the maximum amplitude remains the same, the minimum level is higher than it would have been prior to companding. There is, therefore, an increase in the minimum signal-to-noise ratio at the receiver.

At the receiver, the signal must pass through a complementary circuit (expandor) in order to restore it to its original form (see Figure 2.17). This circuit has the effect that in the absence of any signal the noise will be attenuated more than it would have been without the non-linear circuits, and it has been found that the reduction of noise during periods of silence has the subjective effect of increasing the performance rating of the system.

In practice an instantaneous compandor is not suitable for telephony work, since the phase distortions produced by the transmission link change the actual shape of the transmitted signal, and this will produce inaccuracies at the expander which will have the same effect as an overall non-linear transfer function. For this reason [19] the gain of the compressor circuit is made to vary by a short-term average value of the signal which is unaffected by the phase distortions. This is chosen to be the length of the shortest element of syllabic importance, about 20 ms. Suitable circuits are shown in Figure 2.18. Theoretically the compressed signal occupies a greater bandwidth but this increase is negligible for speech signals with a syllabic time-constant. For speech telephony use, the compressor characteristic which usually has the form $V_{out} = (V_{in})^{1/m}$ with typical values of $m = 2$. Since the compandor increases the average power level of the signal, it is usually necessary to decrease slightly the maximum voltage level so as to keep the average power equal after companding. It has been shown [20] that with a syllabic compandor the subjective improvement produced in situations where such improvement is needed is equal in decibels to ⅔ of the reduction of speech-off noise plus ⅓ of the reduction of the speech-on noise.

A practical syllabic compandor-expandor combination introduces some distortion; also the effect of any gain variations in the transmission path is magnified by the expandor and this can affect the overall stability of a connection. For these reasons, syllabic compandors are used in the telephone network only when they are essential to make an existing circuit acceptable for long distance communication.

If the transmission channel has a perfect phase characteristic and can transmit the increased bandwidth (as in p.c.m. systems), then an instantaneous compandor may be used, as will be discussed in Chapter 10. Instantaneous compandors are an integral part of the transmission system and do not have the disadvantages of syllabic compandors already mentioned. Consequently, there is no objection to their use in a telephone system.

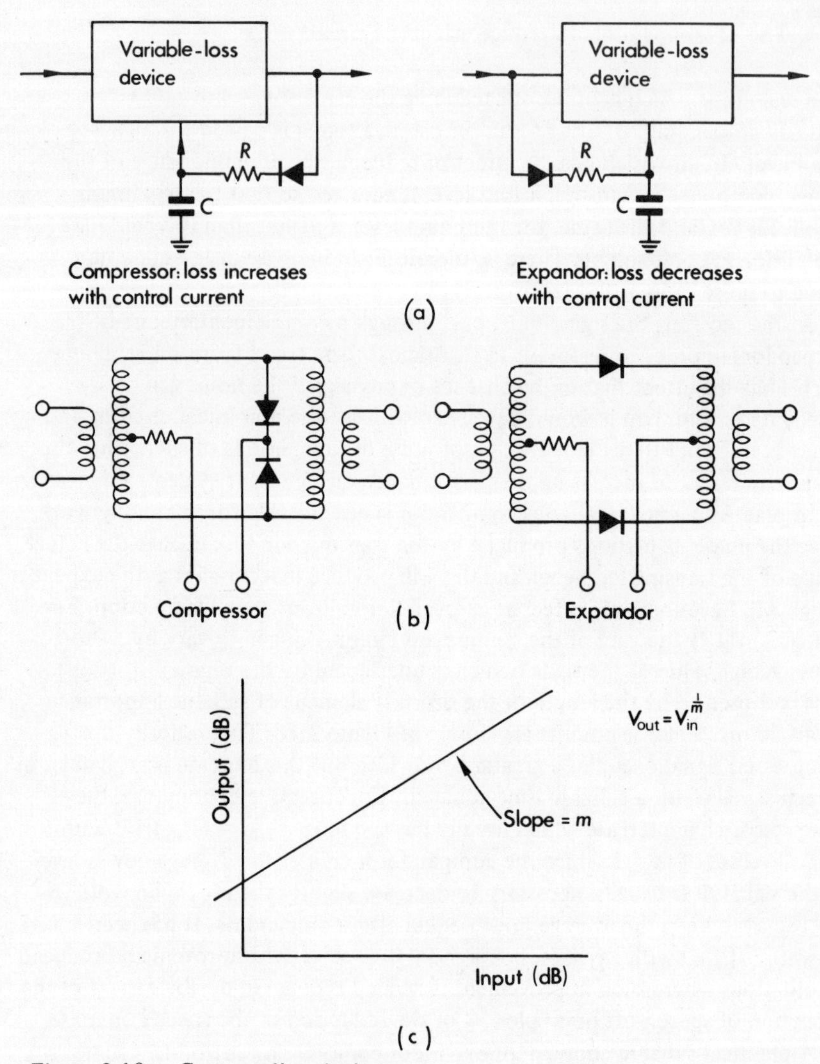

Figure 2.18 Companding devices
(a) Compressor-expandor systems, (b) Actual device (in practice transistors rather than diodes are used) (c) Ideal behaviour

REFERENCES

1. J. S. P. Roberton, 'The Rocking-Armature Receiver', *Post Office Electrical Engineers Journal* **48**, 1956, p. 208.
2. 'Experimental "Electret" Telephone Microphone', *Post Office Telecommunications Journal* **35**, 4, April 1968, pp. 151-3.
3. *C.C.I.T T. White Book III*, Recommendation G121 E.
4. S. A. Campbell and R. M. Foster, 'Maximum Output Networks for Telephone Substation and Repeater Circuits', *Trans. A.I.E.E.* **39**, 1920.
5. H. J. C. Spencer, 'Some Principles of Anti-Side Tone Telephone Circuits', *Post Office Electrical Engineers Journal* **48**, 1955, pp. 208-11.
6. F. E. Williams and F. A. Wilson, 'Design of an Automatic Sensitivity Control for a New Subscriber Telephone Set', *Proc. I.E.E.* **106B**, 1959, pp. 361-71.
7. For example, see discussion in F. A. Wilson, 'Speech Level Control in Telephone Instruments', *Proc. I.E.E.* **114**, July 1967, pp. 907-15.
8. A. H. Flores and T. L. Moore, 'The Evolution of Station Carrier and Recent Operating Experience', *Trans. I.E.E.E.* **COM-19**, April 1971, pp. 211-217.
9. H. J. C. Spencer, 'Optimum Design of Local Twin Telephone Cables with Aluminium Conductors', *Proc. I.E.E.* **116**, April 1969, pp. 481-488. 'A New Method of Local Transmission Planning', *Post Office Electrical Engineers Journal* **63**, 2, 1970, pp. 84-5.
10. 'Local Telephone Networks', *C.C.I.T.T. Handbook*, I.T.U., 1968.
11. See E. A. Guillemin, *Communication Networks* Volume II, Wiley 1935.
12. K. G. T. Bishop and H. E. Robinson, 'Joint Loading of Local-Line Cables', *Post Office Electrical Engineers Journal* **60**, 2, 1967, pp. 101-3.
13. W. T. Palmer, 'Principles of NIC's and the Development of a NIC 2-wire Repeater', *Post Office Electrical Engineers Journal* **51**, October 1958, p. 206.
14. F. B. Llewellyn, 'Some Fundamental Properties of Transmission Systems', *Proc. I.R.E.* **40**, March 1952, pp. 271-83.
15. *C.C.I.T.T. White Book III*, Recommendation G122B.
16. *C.C.I.T.T. White Book III*, Recommendation G131B(a).
17. P. H. Shanks, 'A New Echo Suppressor for Long-Distance Communications', *Post Office Electrical Engineers Journal* **60**, January 1968, pp. 288-92. D. L. Richards and J. Hutter, 'Echo-Suppressors for Telephone Connections Having Long Propagation Time', *Proc. I.E.E.* **116**, June 1969, pp. 955-63.
18. A. W. Thies and R. B. Zmood, 'New Ways of Echo Suppression', *Australian Telecommunications Journal* **1**, November 1967, pp. 14-19. and M. M. Sondhi, 'An Adaptive Echo Suppressor', *B.S.T.J.* **46**, 3, 1967, pp. 497-514.
19. R. O. Carter, 'Theory of Syllabic Compandors', *Proc. I.E.E.* **111**, March 1964.
20. D. L. Richards, 'Transmission Performance Assessment for Telephone Network Planning', *Proc. I.E.E.* **111**, 5, May 1964, pp. 931-40.

Chapter 3

Transmission planning for speech networks

3.1 Aims of a transmission plan

The aim of a transmission plan is to ensure that telephone subscribers can talk to each other with an acceptable standard of clarity and ease, and at a reasonable cost. The derivation of a suitable plan consists of several steps:

(a) Make a precise statement (in statistical terms) of the transmission aim, e.g. that 95 per cent of all calls made by subscribers should be rated satisfactory or better.
(b) Determine what combination of channel degradations are necessary to achieve this aim.
(c) Determine the network configuration and allocate the various degradations to the individual links in an economic manner.
(d) Devise installation rules to achieve the required aims.
(e) Assess in theory and practice whether the network achieves the required transmission aim.
(f) Go back to step (i) and repeat process.

The most economical plan would be to find the minimum acceptable transmission standard and then engineer the network so that no connection is needlessly better than this standard. At the other extreme, the ideal plan is to find the most acceptable standard and then engineer the system so that all connections have this standard. In practice this latter aim is too expensive to achieve in an existing network but most administrations are aiming for it in the long term.

This chapter is concerned with some details of steps (b) to (e) and gives examples of the current practice of the United Kingdom and North American systems, as well as the international transmission plans. At the present time the major design considerations concern loss for national networks and also circuit noise and echo for large national systems and the international system. With the gradual evolution of digital communications these factors will lose their importance and different factors such as quantisation noise, error rate etc. will be the overriding design considerations (this is discussed in the section on pulse code modulation, Chapter 10).

The main reason why loss and echo are major design considerations in a large interconnected network is that, for economic reasons, the subscribers are interconnected, at least at the local level, by 2-wire circuits. As was explained in the previous chapter, there are fundamental limits to the minimum loss of a 2-wire circuit, and once practical problems of parameter variation are also included, this implies that any amplified circuit which must be unconditionally stable for any combination of terminal conditions must be designed for a minimum loss of at least 3 dB each way. Hence each independent link in a 2-wire switched network will add to the loss and this must be carefully controlled. Also the presence of hybrid transformers or some form of 2-wire amplifier will introduce points of reflection for a signal which will be returned to the subscriber as an echo. If the transmission system were 4-wire to the subscriber, then the echoes would be eliminated, and it would be straightforward to have a small fixed loss for the system. The use of a 4-wire system would not produce any advantages for circuit noise design.

Another major design factor, especially for long distance networks, is to ensure that the intelligible and non-intelligible cross-talk is kept to adequately low levels.

3.2 Interconnection techniques for transmission systems

In a civilian telephone network the subscribers are connected to a local exchange by 2-wire circuits and these exchanges are interconnected via a hierarchy of other exchanges. Military systems frequently have a different structure in order to make them more flexible and more tolerant to damage, but these are not considered here. There are basically four techniques, from the transmission viewpoint, used in a multi-link switched connection.

(a) *Simple 2-wire switching.* This is the most straightforward technique whereby the individual links are interconnected by a 2-wire switch. The individual links are completely independent and have a loss depending upon their length and the type of cable, loading units etc. If some form of amplification is provided this may be on a 2-wire or a 4-wire basis within the link but, whatever amplification technique or combination of techniques is used, the circuit is presented to the switch in a 2-wire form. If there is some amplification then there is a minimum loss that the circuit can operate under; this depends upon the tolerances of the different components, but is normally of the order 3 to 6 dB. There is therefore a limit to the number of links which may be interconnected in this manner. For instance, if each link operates at a 3 dB loss then an n-link connection will have a loss of $3n$ dB.

(b) *2-wire switching with pad switching.* In the long-distance network it is possible and economic to control the impedances of the individual links as seen from the two ends. If an amplified 2-wire circuit has only to be stable over a limited range of terminal impedances then, as was explained in the previous chapter, it is possible to provide stable operation at zero loss or even at a gain, provided that at the internal interfaces the links present well defined impedances to each other. Hence several 2-wire circuits may be interconnected and the fundamental gain limitation only applies to the overall connection rather than each circuit. However, in a switched network an individual link may be used singly or in a multi-link interconnection and the link must be stable under all uses. To provide for this it is necessary to have a matched fixed attenuator

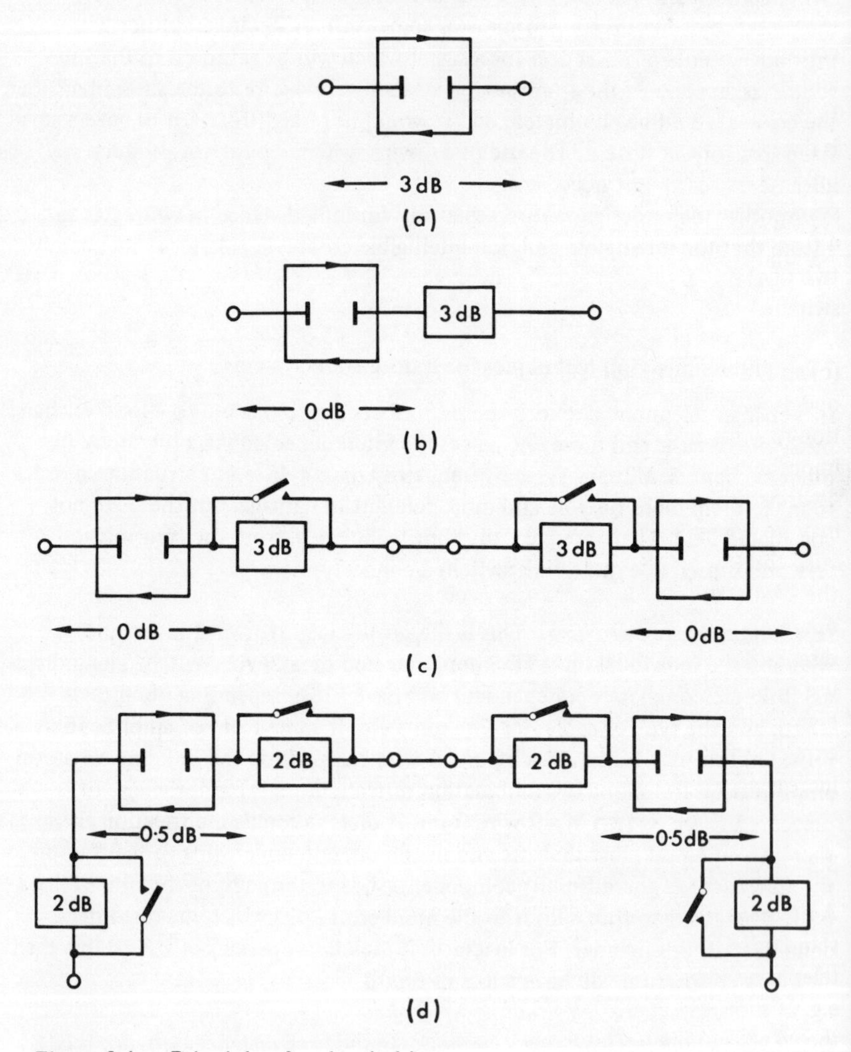

Figure 3.1 Principle of pad-switching
(a) Stable 2-wire amplified circuit, (b) 2-wire amplified circuit with identical stability to (a), (c) Principle of pad-switching, (d) A more practical form of (c)

(called a pad) connected to each circuit, which may be switched out when the circuit is used in a multi-link connection.

The way this may be achieved is shown in Figure 3.1. Assume, for example, that an individual link is in fact provided on a 4-wire amplified basis and the expected range of the balance return losses and the amplifier stability is such that the circuit must operate at a 3 dB loss, as shown in Figure 3.1(a). In practice the same stability may be obtained if the 4-wire portion of the circuit is operated at 0 dB and a 3 dB attenuator (matched to the line impedance) is added at one end, as shown in Figure 3.1(b). If two of these links are now connected in tandem, then it is possible to switch out one of the attenuators, and the two-link circuit may then operate at 3 dB loss as shown in Figure 3.1(b). In practice[1] it is not as simple as this since for each link added to the chain it is necessary to operate the overall connections at a slightly higher loss in order to provide for the larger spread of parameters in a multi-link call. A practical system is to operate at $(4 + 0{\cdot}5n)$ dB, i.e. a single link operates at 3·5 dB, two at 4·0 dB overall etc. This may be achieved as shown in Figure 3.1(d). Whenever two of these circuits are interconnected, the 2 dB pads at the interface are switched out.

(c) and (d) *4-wire switching with and without pad-switching.* If the switch can operate on a 4-wire basis rather than 2-wire, then a low-loss multi-link connection may be made more reliably and without the necessity to control the line impedances which there is in 2-wire with pad-switching. The advantages of 4-wire switching are that, at the expense of a larger switch, a lower overall loss may be generally achieved and without the necessity for impedance control. Also there are no internal echo paths present, which can present problems for the 2-wire pad-switching system. Each link in a 4-wire system must in practice operate a slight loss in order to keep the complete connection stable (this is discussed in more detail in section 3.4) and this may be achieved either by the use of pad-switching similar to that in 2-wire switches or by placing a sufficiently high fixed loss in the terminating links to keep the connection stable up to an expected maximum number of links in tandem. Both systems are coming into increasing use in long distance and international use.

3.3 National transmission plans

A national transmission plan is based on the use of a national transmission standard. This is a representative transmission system consisting of two local telephone circuits (specified handset, limiting subscribers line and feeding bridge, e.g. as shown in Figure 3.2). There are connected by an attenuator, X, simulating the effect of junction and trunk losses. For instance, for the circuit shown, it is found that[2], in the absence of circuit noise but in the presence of a representative ambient room noise, an attenuator setting of $X = 19$ dB, 90 per cent of subjects will give the connection a voting of 'fair' or better.

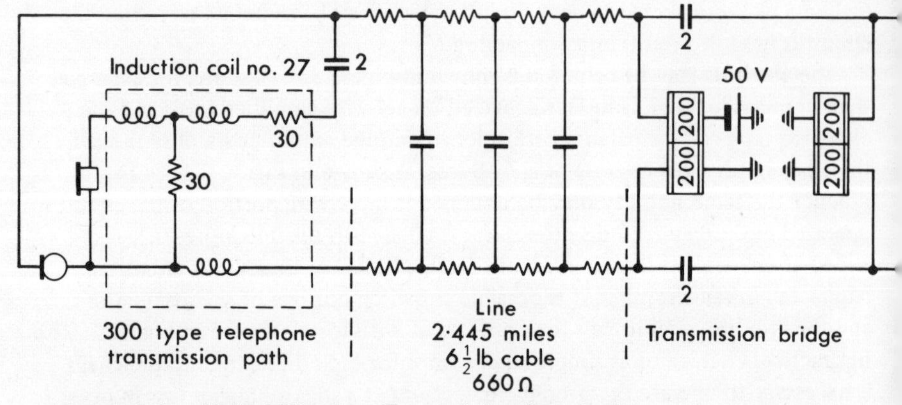

Figure 3.2 United Kingdom Transmission Standard

If now some part of the standard local telephone circuit is replaced by an improved component, the attenuator setting may be found for which 90 per cent of people would give a 'fair' or better rating, and the difference in settings give the additional attenuation that the system may introduce. In an existing system, this addition can only be utilised on the subscribers line if the exchange has to continue to use the older components as well.

The use of the word standard in this connection is probably misleading since the circuit is a limiting case and in fact nearly all the exchange connections will be better than the 'standard'.

The United Kingdom Plan

The junction and trunk network can now be planned such that X is not exceeded (the British Post Office use a figure of 20 dB). Civilian telephone networks are based on a hierarchy of switching centres as shown in Figure 3.3 [3]. A subscriber is connected either to a local exchange (so called 'minor') or else to a dependent exchange parented on to a minor exchange. There are then various inter-exchange junctions if the traffic justifies the route, or else a single tandem exchange. If a direct route is not available then the call is routed to a group switching centre (GSC) and thence to a remote GSC or else direct to the remote exchange. There are about 400 GSC's in the United Kingdom and these are obviously not all fully interconnected and a call may have to go via an intermediate GSC. If no such route is available, then (until recently) the call would have to be switched via the highest rank of zone switching centres of which there are twenty-four existing or planned. These zone centres are nearly all fully interconnected but the exceptional call might need to go through an intermediate zone centre. The nominal transmission losses for the British system are shown in Figure 3.3 and it may be seen that the maximum transmission loss for a

dependent-minor-group-zone-zone-zone-group-minor-dependent

is 15 dB. However, this does not include the losses introduced by the switching

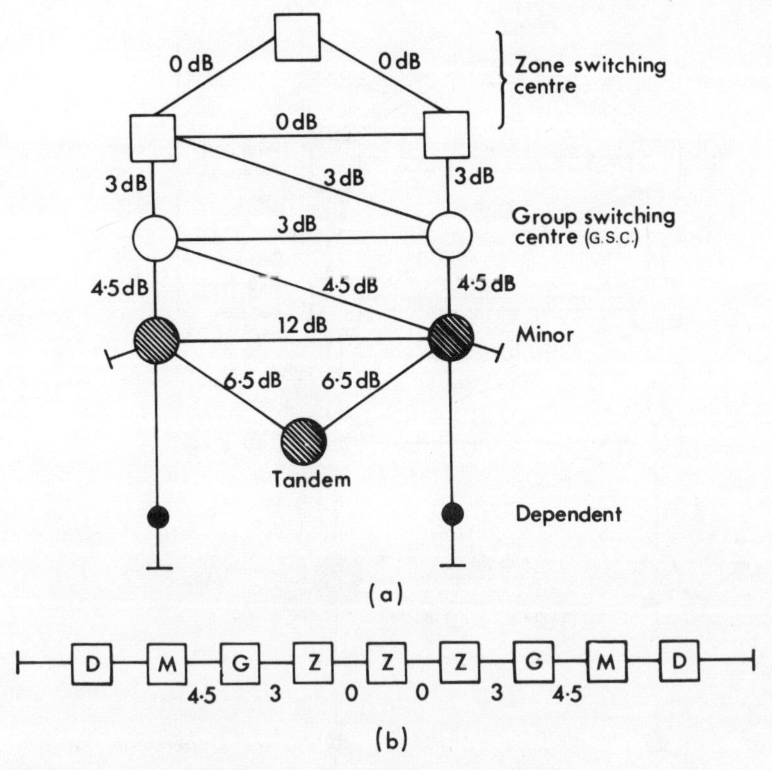

Figure 3.3 United Kingdom Transmission Plan (1933)
(a) Basic switching system (Old Plan – 1933), (b) Maximally adverse connection

and signalling systems at the intermediate exchanges, which amount to about 1·5 dB per exchange on average. Hence, under these most adverse conditions with seven intermediate exchanges, the total transmission loss is 25·5 dB which leads to a low probability of satisfactory connection, and even this target was not achieved in practice. Until the advent of subscriber trunk dialling (s.t.d.) this could be obviated by the operator testing the circuit before connecting the subscriber and obtaining an alternative if necessary. However, for long distance s.t.d. a much better transmission plan is necessary and a new one was introduced in 1960, as shown in Figure 3.4. Another reason for introducing the new network is that it is designed to permit a much faster setting-up time than the old GSC's and zone centres. This is essential if s.t.d. is to be used for multi-link calls. This new network is based on a trunk transit network which is 4-wire switched and which would take the extra-long-distance calls for which the old network was unsuitable. The transit network is designed to have a 7 dB loss independent of the number of intermediate switching stages and this is possible only with the use of a completely 4-wire switched system. The nominal transmission loss is now only 19 dB and it is estimated [4] that in practice a

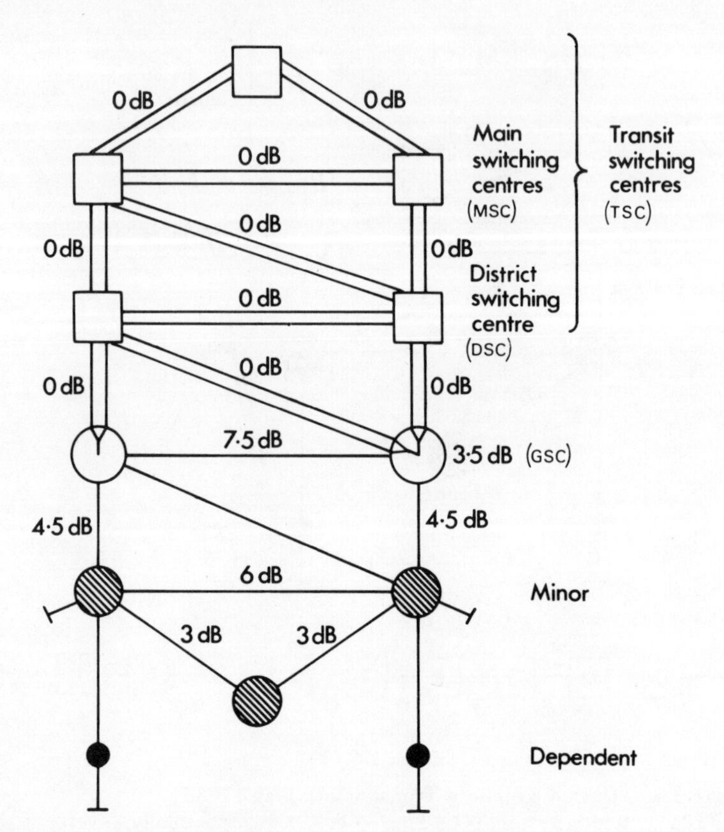

Figure 3.4 New British Transmission Plan (1960)

mean of 19·5 dB (σ = 3 dB) will be achieved. There will be forty-two of these transit switching centres (TSC's) and six of them will be fully interconnected so as to reduce the maximum number in tandem to five.

The figure of 7 dB gives an adequate margin of stability for the maximum number of 4-wire circuits that are likely to be connected in tandem. A fixed value, rather than a $(4 + 0{\cdot}5n)$ dB type of plan, is preferable as it is easier to set-up and test and, more importantly, it ensures that the signalling receivers used at transit exchanges have the same nominal loss to work over.

North American Transmission Plan

The transmission plan for a very large country such as North America or Australia differs from that of smaller European countries in several respects. The most important are:

(a) There are more ranks in the hierarchy of exchanges so that the maximum number of links on a connection is potentially greater.

(b) Echo is a real problem.

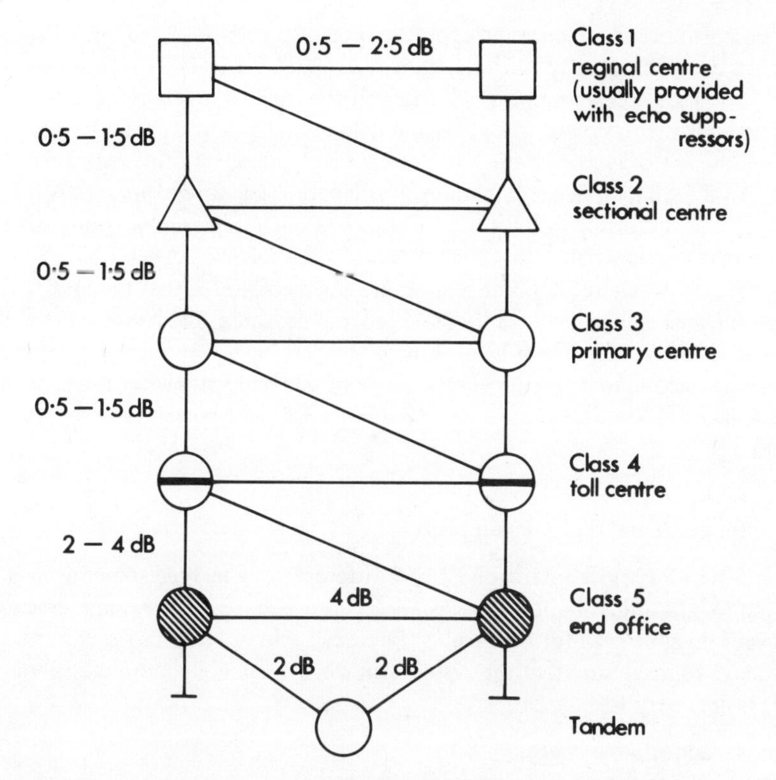

Figure 3.5 North American Transmission Plan. The basic plan is $T = 4 + 0{\cdot}5n + Xl$ dB where n is the no. of circuits, X is the Via Net Loss Factor, l is the length

Notes

1. There is 4-wire switching at the regional centres and the majority of sectional centres. Future plans involve extension of 4-wire switching to some Class 4 offices.
2. There may be direct routes between Class 1 and Classes 3 and 4 between Class 2 and Class 4 offices. Also end offices may be parented on to Class 3 offices.

The current approach of North America[5] with their terminology is outlined in Figure 3.5. There is 4-wire switching at the regional centres and at the majority of sectional centres. 2-wire switching is used elsewhere in the higher network, but considerable care is taken to control the impedances at the exchanges. Echo suppressors are usually fitted on the regional centre to regional centre links. It may be seen that there are much lower losses on the trunks in the long distance network. The values chosen depend not only upon the type of trunk but also upon the delay time it introduces. The loss between the two end offices (i.e. local exchanges) is called the Via Net Loss (VNL) and it takes the form

$$T = 4 + 0{\cdot}5n + Xl \text{ dB}$$

where n is the number of links, l is the length of the connection and X is the Via

Net Loss Factor, which depends upon the type of transmission medium. For instance, for high velocity carrier systems the value of X is 0·0015 dB/mile, so for a 3000 mile connection this will add a further 4·5 dB to Via Net Loss[6].

The implementation of this plan originally used 2-wire pad-switching to achieve the $4 + 0{\cdot}5n$ part of the Via Net Loss and individual trunks are lined up to 0·5 to 2·5 dB loss, depending upon their length. The current practice is to make all the end office to toll centre trunks work at a minimum loss of 2 dB and ensure that the impedance that they present to the toll exchange is controlled.

A further difference between a large and small country is that the local telephone area often covers a very large geographical area, and hence there is the possibility of a wide range of lengths of subscribers lines. This normally implies that some techniques for reducing the range of received volumes at the exchange are necessary. For instance, the use of loading coils and booster batteries on long local lines.

3.4 International transmission plans

When it is necessary to interconnect two different systems then some form of international standardisation is necessary to ensure that effective communication is always possible even for the most difficult call routing. It is also necessary nowadays to make sure that the system can work completely automatically.

It is necessary to:

(a) allocate the losses,
(b) define the sending and receiving levels,
(c) ensure that stability conditions are met,
(d) allocate noise contributions economically,
(e) allocate propagation time,
(f) eliminate objectionable echoes,
(g) define other impairments.

These are all achieved via CCITT recommendations.

The standardisation is performed with the aid of a reference system held at the ITU laboratories in Geneva. This system is defined and calibrated in terms of fundamental physical units so that it may be reproduced and checked if necessary and it goes by the name NOSFER (Nouveau Système Fondamental pour la détermination de Equivalents de Référence)[7].

History of NOSFER

In order to understand the reasons for the particular configuration some of the history of the standard systems[8] must be known. The original equipment installed in 1928, was known as SFERT (Système Fondamental Européen de Référence pour la Transmission Téléphonique). This was designed to be (for its day) a high quality speech system and it used standard components with known sensitivities and frequency responses. The equipment was used to provide a loudness balance measurement for test systems. However, as the quality of commercial telephone sets improved with time it was found that volume was not an adequate indicator of performance, since a better quality at the same volume would give high intelligibility.

In order to measure intelligibility, a new equipment was introduced in 1949 called AREAN (Apparatus de Référence pour la détermination des Affaiblissements pour la Netteté–Reference apparatus for the determination of transmission performance ratings). This equipment was originally derived from the British Post Office and had been designed to be equivalent to a 'metre air path' when the speaker was a fixed distance from the microphone and the listener was closely coupled with the earphone[9]. The attenuator setting in this condition was 30 dB and the frequency response was adjusted to take account of the obstructing effect of the listener's head in a free field environment.

In order to conduct articulation tests the AREAN equipment was modified by the addition of a band-pass filter which was designed to simulate the typical attenuation/frequency distortion produced by the tandem connection of twelve carrier channels. A fixed level of background noise was also injected into the system. This noise was designed to simulate the effect of room noise rather than circuit noise and was injected electrically rather than acoustically since this was found to be more reproducable. The spectrum of the noise was tailored to be that found at an average telephone location ('Hoth' spectrum)[10]. The AREAN with these modifications was then called SREAN (système de Référence etc.) and was used by sending a series of Esperanto sounds chosen at random and measuring the percentage of sounds correctly understood. The SREAN was then used in what is called an indirect balancing manner to obtain ratings for test system. This consists of taking a series of intelligibility tests for different attenuator settings of the standard system and again for the test system. These are then plotted as shown in Figure 3.6 and the difference in

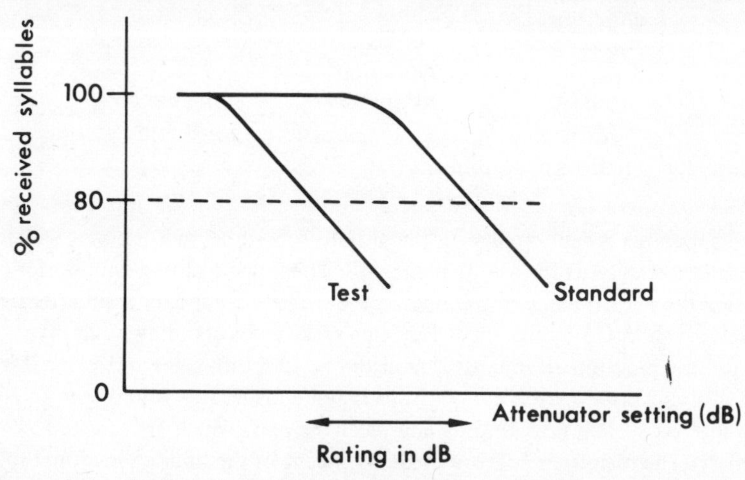

Figure 3.6 Indirect balancing technique for articulation measurements

attenuator settings at the 80 per cent reception level is taken as a measure of the articulation efficiency of the system. However, as improved modern telephones have come into more universal use, the degradations in transmission due to distortion and noise have become relatively less important and the most significant factor has been found to be loss. For this reason this technique is rarely used today.

In 1956 it was decided to revert to volume measurements but by this time the old SFERT equipment was very old. It was decided therefore to modify the AREAN equipment to reproduce the frequency response of SFERT (which was nominally high-quality but in fact dropped slowly after about 4 kHz). With these modifications the AREAN was called NOSFER and this is where the story stands at present.

Hence NOSFER has an arbitrary but fairly reasonable frequency response and its sensitivity is such that it produces a signal of equal loudness to a 1 metre air path when its attenuator is set to 33 dB.

NOSFER equipment

The system consists of a moving coil microphone, send and receive amplifiers and a moving coil receiver of well defined characteristics. A standard speech volume measuring set is connected across the output of the sending end (see Figure 3.7).

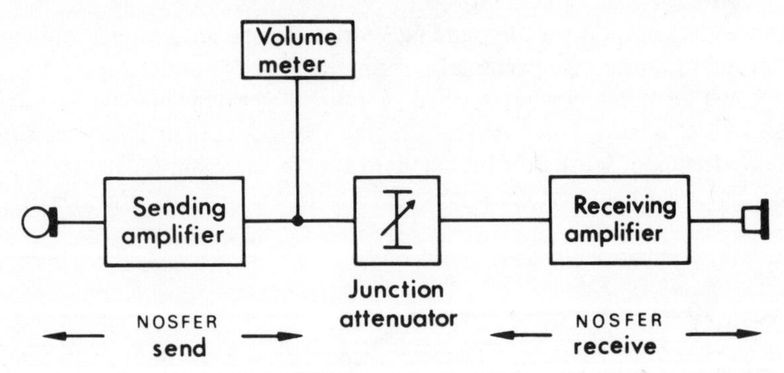

Figure 3.7 NOSFER Equipment

The apparatus is calibrated with the aid of probe-tube microphones which act as secondary standards for use with an artificial ear and a closed coupler for observing the performance of the microphone. These secondary standards are themselves calibrated by means of Rayleigh disks and a standing wave tube.

In use the junction attenuator is adjusted to 15-24 dB since this is the level at which most accurate speech comparisons may be made. It is found convenient to divide the test system into sending and receiving parts which are assessed separately. The measurements are made by means of the 'hidden loss' method as, for instance, is shown in Figure 3.8(b). In this method, an operator speaks

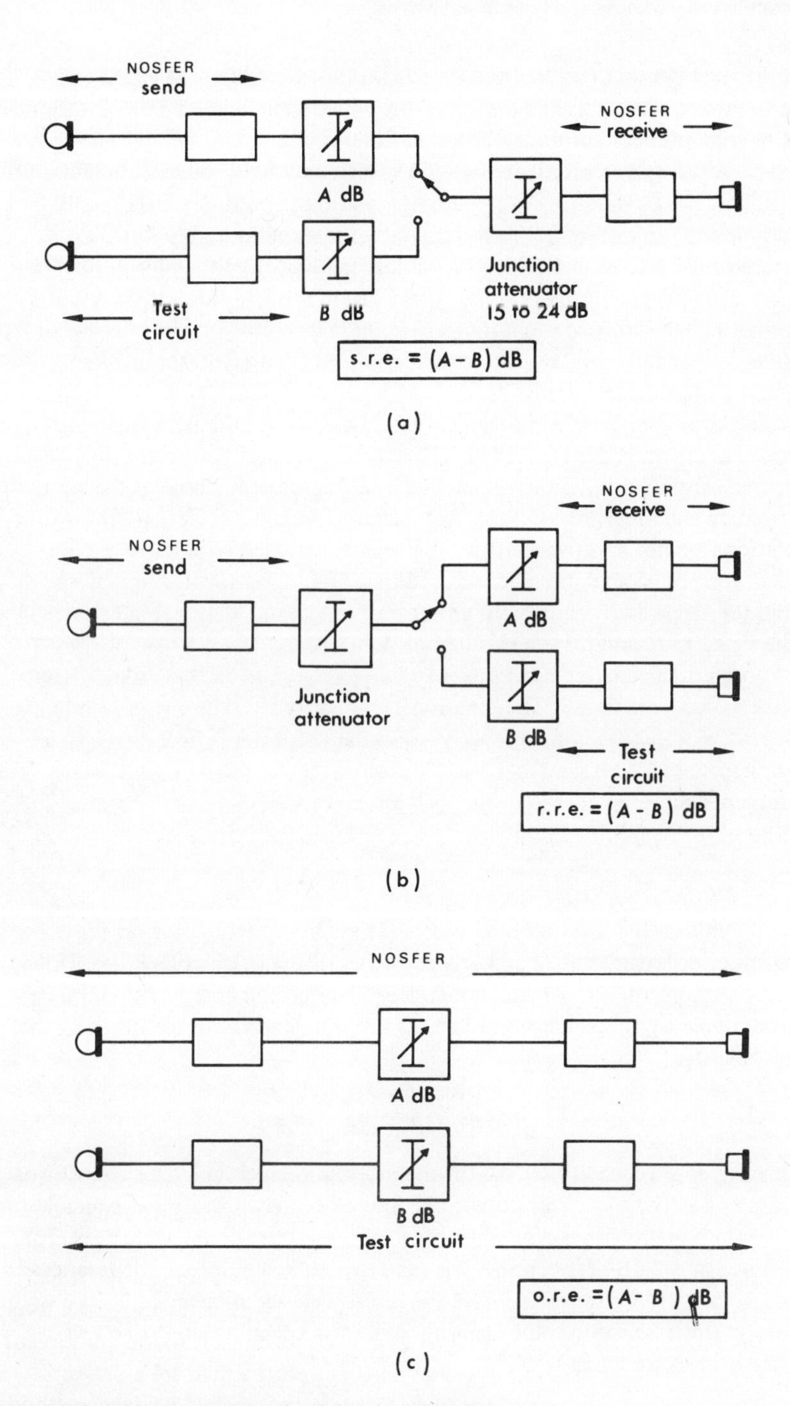

Figure 3.8 Measurement of reference equivalents
(a) sending, (b) receiving, (c) overall

into the send circuit at a fixed distance from the microphone (a guard ring is used to ensure this) at such a level that the volume reading has a pre-determined reading. She places an arbitrary loss B in the test circuit and switches between the two. Another operator listens to the speech coming through both channels and adjusts the attenuator A until the signal through both channels sounds equally loud. The difference between the two attenuations $A - B$ is then a measure of the relative efficiency of the test sending circuit, and is defined as the *sending reference equivalent (s.r.e.)*. If the attenuation in the test circuit is less than that of the standard then the test circuit is less efficient and this gives a positive reference equivalent. Thus the definition is purely in terms of an arbitrarily physical standard.

The *receiving reference equivalent (r.r.e.)* may be found in an analogous manner, as shown in Figure 3.8(b). The overall reference equivalent may also be measured as shown in Figure 3.8(c); this is not necessarily equal to the sum of the sending and receiving reference equivalents, since the effect of band-width limitations are not additive, but this discrepancy is small with NOSFER on modern telephone apparatus. It is found that on international connections, the overall reference equivalent (with an error of less than 1 dB) is equal to the sum of the s.r.e., r.r.e. and the 800 Hz loss of the circuit.

National transmission standards may be compared to NOSFER and their relative equivalents found. These national standards can then act as secondary standards. For instance, the reference equivalents for the British Post Office limiting line are 12 dB send and 1 dB receive. Other countries use different instruments and their figures differ. For instance, Austria has a 7 dB s.r.e. and 4 dB r.r.e. for its national standard.

The 1964 International Transmission Plan

Until the mid-sixties international calls were nearly all connected on a 2-wire basis under operator control and few countries offered transit facilities. Under these conditions the CCITT recommended[11] that the maximum reference equivalent for a connection should be 40 dB overall and with the limits of the national system,

18·2 dB s.r.e.
13·0 dB r.r.e.

i.e. leaving approximately 8·8 dB for the international circuit. This circuit was recommended to have a 7 dB loss so the approximately 1·8 dB was available for bandwidth impairment and switching loss of the international exchanges. The difference of 5·2 dB between the s.r.e. and r.r.e. reflected typical differences for handsets in common use in the days the limits were recommended

In the United Kingdom the limiting values for a handset, long line and feeding bridge, are taken to be

12 dB RE s.r.e.
1 dB RE r.r.e.

Hence to meet these recommendations there must be a maximum of 6·2 dB loss on the send side and 12 dB on the receive side. Since, in general, the loss in either direction is made nominally equal, then it is the 6·2 dB which is the limiting case. Examination of the old U.K. plan shows that for provincial subscribers it was not possible to keep within this limit if the international exchanges were of zone switching status, and special high quality circuits to the provinces were necessary for international use.

For a world-wide, fully-automatic, subscriber-dialled service, this plan was inadequate and, after much discussion, in 1964 the CCITT recommended a new plan based on a hierarchy of 4-wire switched international exchanges[12].

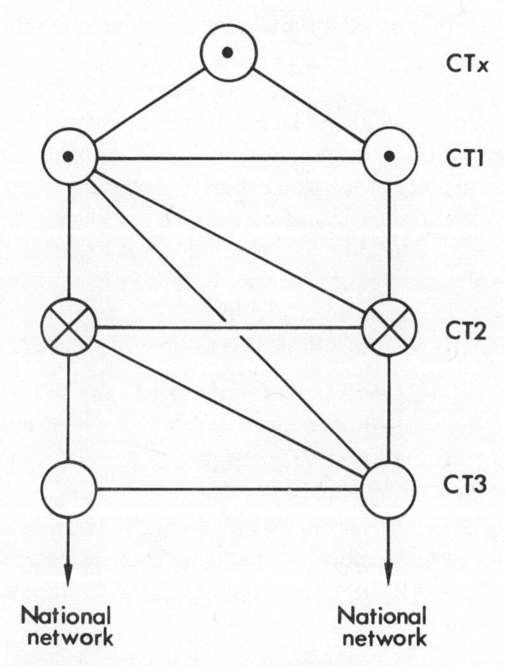

Figure 3.9 The international exchange system

These are in three levels, designated CT1, CT2, and CT3, as shown in Figure 3.9 (CT stands for Centre du Transit). The CT1s are the highest level and are intended to be fully interconnected, although in exceptional cases it is envisaged that an additional transit centre may be needed of unspecified rank (called CT*x*). At present there are seven CT1s planned to cover the world. These will be in London, New York, Moscow, Sydney, Tokyo, Singapore and somewhere in India or Pakistan. The principle countries in each CT1 zone have been assigned one or more CT2s depending upon their size (e.g. there is one in the United Kingdom, twelve in the United States and two in Canada). The CT3s are at the interface between the national and international network. Hence, when this

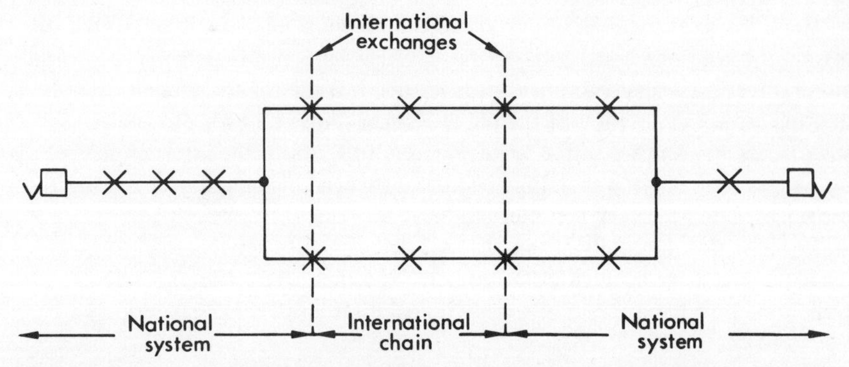

Figure 3.10 Definition of constituent parts of an international connection

system is fully implemented, it will be possible to interconnect any two international exchanges with a maximum of five international circuits (or exceptionally six). A further feature of the plan is that it is possible to extend the 4-wire international circuit into the national network.

For international systems, the first requirement is to define the constituent parts of the system so as to be able to specify interface requirements. The terms used by the CCITT are illustrated in Figure 3.10. A complete *international telephone connection* consists of three parts.

(a) *An international chain* made up of one or more 4-wire international *circuits*. These are interconnected on a 4-wire basis in the international transit centres and are also connected on a 4-wire basis to national systems in the international centre.

(b) *Two National systems*, one at either end. These may comprise of one or more 4-wire national trunk circuits with 4-wire interconnection, as well as circuits with 2-wire connection up to the terminal exchange and to the subscribers.

Virtual switching points

In international 4-wire circuits it is necessary to specify the level at which they are assumed to be interconnected. This is done by defining *virtual switching points* which are theoretical points with specified relative levels at which switching is assumed to be performed. The values chosen by the CCITT are:

–3·5 dBr sending
–4·0 dBr receiving

which implies a 0·5 dB nominal loss on the international circuit. For an isolated 4-wire circuit the relative levels are defined relative to the 2-wire input of a hypothetical hybrid transformer that would be present if the system were terminated at either end. This is illustrated in Figure 3.11(a). The –3·5 dBr is

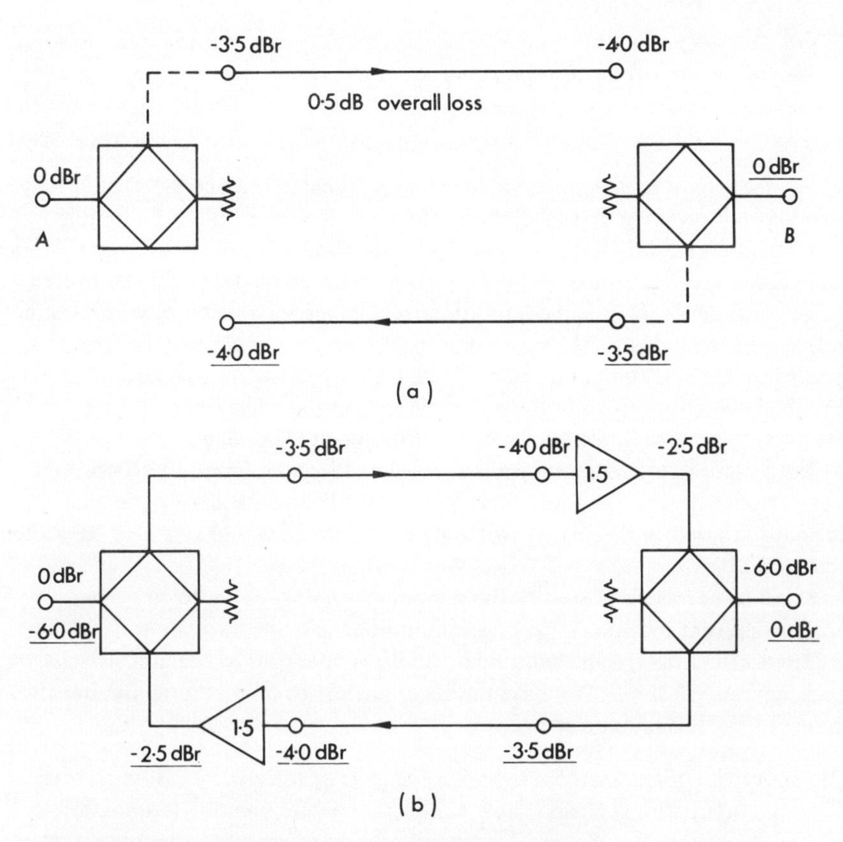

Figure 3.11 Definition of virtual switching points
(a) Definition of relative level on 4-wire circuit (b) Use of a 4-wire circuit to give a 6 dB loss 2-wire circuit. All underlined levels refer to point *B*

relative to a point on side *A* and –4·0 dBr is relative to a point on side *B*. If it were desired to use the international circuit to make a 6 dB 2-wire circuit, then two 1·5 dB amplifiers would be needed as shown in Figure 3.11(b).

For the case where there is more than one international circuit in tandem, then each will have been designed to work to a virtual switching point of –3·5 dBr send and –4·0 dBr receive relative to the reference points of the *circuit*. The reference point of a *connection* is usually taken as the entry to the trunk network within the national system, and hence in a particular connection the actual relative levels at which the individual circuits operate depend upon their position in the chain. (Note that since the units all operate at a 0·5 dB loss, the design level is in fact an upper limit).

The actual level at which switching takes place and the level which is presented to and expected from the international circuit is a matter for the individual administrations to agree amongst themselves. The use of switching points only allows for a common basis of specification.

For the new system the recommended maximum limits for the reference equivalents at the virtual switching points are

20·8 dB s.r.e.
12·2 dB r.r.e.

for 97 per cent of actual incoming or outgoing calls [13]. Since there will be a maximum of six international circuits, each with a nominal 0·5 dB loss, this leads to an overall reference equivalent of approximately 36 dB. Although this value is still high for a modern network (the preferred range is 12±6 dB overall reference equivalent), it is the limit of current economic feasibility with existing national systems. Also, the upper limit of 36 dB o.r.e. should only be met on a small proportion of the actual calls. The difference between the s.r.e. and r.r.e. is 8·6 dB rather than 5·2 dB in the old recommendation which takes account of the relative increase in receive efficiency of modern telephones.

Note that the reference equivalents for the old plan referred to a 2-wire international circuit. They were in effect quoted for a 0 dBr level. The old standard referred to the virtual switching points which would give 21·7 dB s.r.e. and 16·5 dB r.r.e.

It is also necessary to specify the minimum sending reference equivalents in order to prevent over-loading of the international circuits by operators at the international exchange and subscribers on short lines near to the international exchange. As yet the CCITT have not recommended a minimum figure, but the value of 6·5 dB is under discussion.

Characteristics of National Systems forming part of International Connections
Since the international chain is now completely 4-wire then the termination is part of the national system and hence will have a considerable effect upon the stability and echo properties of the international chain [14]. The national portion of an international connection is shown in Figure 3.12(a), and it is the transmission path *a-t-b* which will affect the 4-wire circuit where *a* and *b* are the virtual switching points (i.e. the sum of the losses *a-t*, *t-b* and the balance return loss at the terminating set). Calculations similar to those in Chapter 2 show that an adequate stability margin may be obtained if the loss *a-t-b* has a mean value of at least $(10 + n)$ dB. Hence, if the balance return loss has a mean value of 6 dB then *a-t* and *t-b* must be at least $(4 + n)$ dB, i.e. $(2 + 0{\cdot}5n)$ dB each if no differential gain is used. This may be achieved directly as shown in Figure 3.12(b) by the use of 4-wire pad-switching in the national circuit. An alternative approach is that used in the United Kingdom which is to make the losses equal to 3·5 dB in each direction for all cases, since there is a maximum of three national circuits to the termination sets. The individual circuits may now be used at 0 dB loss and 3·5 dB added in the termination set. This implies that the trunk transit network may be used as it stands for extending the international call to the nearest group switching centre. Even under these conditions there will be a small proportion of calls not meeting the CCITT reference equivalent recommendations, but the ultimate aim is to meet them for 100 per cent of all connections.

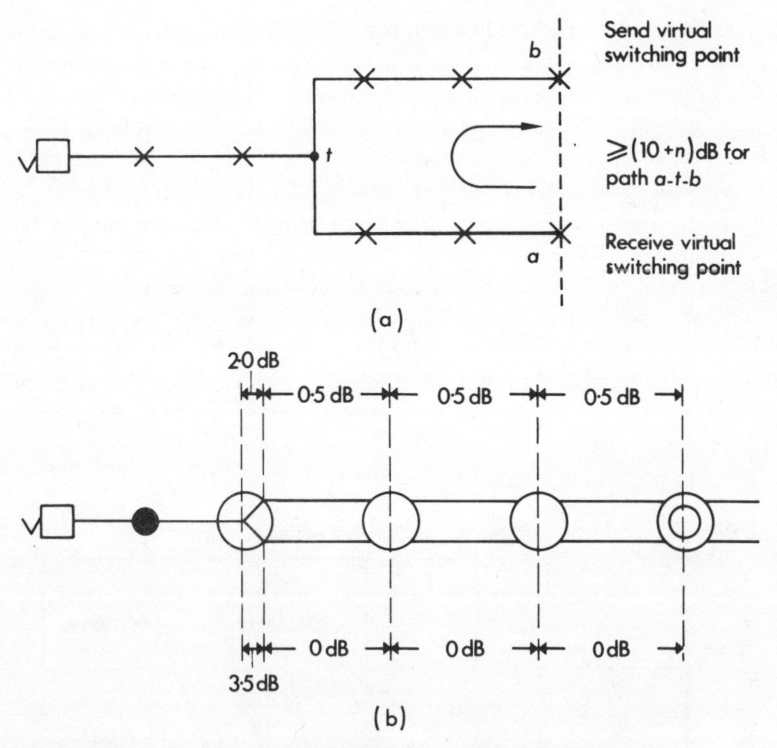

Figure 3.12 Stability requirements for national systems
(a) Loop formed by national system on international connection
(b) Alternative transmission plans to satisfy stability requirements

In a similar manner there must be a limit on the mean loss over the speech band to provide adequate echo attenuation. In particular a mean value of $(15 + n)$ dB is recommended for a national circuit which does not introduce any significant delay. If the national network does introduce significant delay (i.e. greater than 10 ms one-way propagation time to the termination set) then either the transmission loss must be increased or else echo suppressors fitted.

3.5 Noise allocation in large systems

After loss and echo, the next most significant channel degradation is noise. There are four main classes of noise:

(a) white noise,
(b) impulsive noise produced by interference,
(c) power hum picked up from nearby mains cables,
(d) cross-talk from other circuits.

These all have different subjective effects and affect the various services in different ways. Impulsive noise is very difficult to even measure reliably, let alone specify. Power hum is only a significant problem on local lines where they may run near to mains supplies. The usual criterion is to define a maximum limit (e.g. 1 mV) that appears at subscribers due to the local line alone.

The problem of allocation of the white noise in a large system from the telephonic viewpoint is amenable to some analysis [15,16]. The first necessity is to find experimentally what the user's reaction to the background noise is. This is found to vary according to the received signal level, but not in any easily calculable manner. Hence it is necessary to conduct conversational tests over a range of different received noise powers and received speech volumes. Some typical results are shown in Figure 3.13 where the noise power levels (psophometrically weighted) are referred to the 0 dB r.e. receive point, i.e. the noise levels are quoted independently of the received speech volumes. For instance, for a noise level of –60 dBmp then for a 24 dB r.e. the level of dissatisfaction is 2·3 per cent, but for a noise level of –68 dBmp and a 32 dB r.e. (i.e. the same signal to noise ratio) the dissatisfaction is 10·1 per cent.

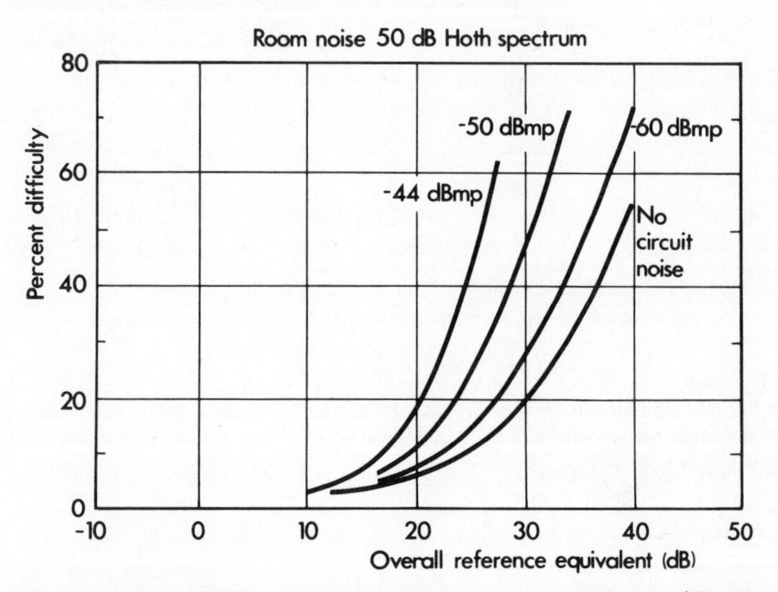

Figure 3.13 Effect of circuit noise on conversational difficulty (Circuit noise referred to 0 dB RE receive point)

The problem of noise allocation in a practical system is solved in an iterative manner by first making an economic allocation and then computing what percentage of users will find the system unsatisfactory. Since there are an infinite number of possible connections, the procedure used extensively for noise studies is that of the *hypothetical reference connection* (h.r.c) [17]. These are theoretical models whereby the average and maximum noise powers contributed by the circuits and exchanges may be specified. There are many different h.r.c.'s in common use for different purposes. A typical one is shown in Figure 3.14 and is the longest international connection envisaged by the CCITT.

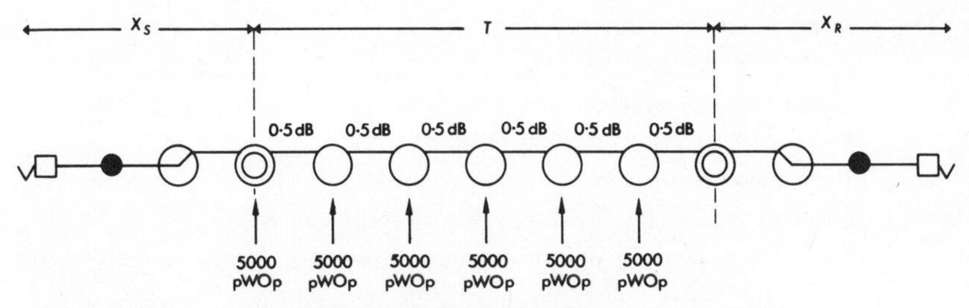

Figure 3.14 A Simplified hypothetical reference connection
X_S is the reference equivalent on the send side
T is the transmission loss of the international chain
X_R is the reference equivalent on the receive side
Noise powers refer to the 0 dBr point of their *own* circuit

This shows the nominal losses together with their variations. Suitable noise allocations may now be made for each component of the connection and these are also shown in Figure 3.14. The noise powers are all expressed in pW0p, i.e. referred to the 0 dBr of the component, i.e. in the case of the 4-wire international circuits the noise contribution of each circuit is referred to its own zero reference point.

It is now possible to compute the total noise power referred to a single reference point of the connection, e.g. the 0 dBr point of the first international circuit in the sending chain. For instance, the mean noise power of the second circuit is 5000 pW but this is at a level –0.5 dB relative to the input of the first circuit. Hence, referring this power to the input of the first circuit will make it 5600 pW. Added to the 5000 pW due to the first circuit, this gives 10 600 pW for the first two circuits. The next stage is to compute the probability of a given receive and send reference equivalent. The subjective tests then give a measure of the probability of dissatisfaction for these conditions. The final step is to compute the overall probability of dissatisfaction for the reference connection. If this is not low enough then the process must be repeated using different allocations of noise to component in the connection.

In practice the actual statistical calculations are somewhat more complex than outlined above, and the details may be found in Richards' paper.[15] On the basis of this type of computation the CCITT have recommended that the noise contribution of the international chain, when referred to the zero reference point of the first circuit in the chain should be 50 000 pW0p (–43 dBm0p)[18]. It has been calculated that with this noise limit, the percentage of connections deemed unsatisfactory will be 10 per cent or less. With less complex routings the noise and hence the dissatisfaction will be less.

The final stage is to allocate this total noise contribution to the individual links. For long distance transmission systems the noise power is proportional to length, and, for instance, the CCITT recommendations are[19]

(a) *Short-distance circuits*
Maximum value 2000 pW, preferably 500 pW.

(b) *Long-distance 2500 km*
Cable or radio-relay system 3 pW/km.
Open-wire carrier systems 7 pW/km.
+ 2300 pW for multiplexing equipment.

(c) *Very long-distance 25000 km*
Cable or radio-relay 3 pW/km or preferably 2 pW/km.
Submarine cable 3 pW/km on worst channel, 1 pW/km mean.
Radio systems 10 000 pW.

In fact since the systems have to take data and telegraph signals in addition to speech then other noise objectives are usually specified, e.g. for a 2500 km circuit the recommendations are:

(a) *Hourly mean requirement* – mean psophometric noise power shall not exceed 10 000 pW during any hour. This hour should be any busy hour for a cable system or for a radio system any hour during the worst fading season

(b) *One minute mean requirement* – mean psophometric noise shall not exceed 10 000 pW for more than 20 per cent of any month or 50 000 pW for more than 0·1 per cent of any month. This requirement is necessary to cover radio-relay systems.

(c) *Five minute mean requirement* – the unweighted noise power with five minutes time constant shall not exceed 10^6 pW for more than 0·01 per cent of any month (4·3 min). This figure is satisfactory for 50 baud f.m. telegraph (see Chapter 11).

REFERENCES

1. H. R. Huntley, 'Transmission Design of Inter-Toll Telephone Trunks,' *B.S.T.J.* **32**, September 1953, pp. 1019-36.
2. E. W. Anderson, 'Local Networks', *Proc. I.E.E.* **111**, 4, April 1964, pp. 713-26.
3. W. J. C. Tobin and J. Stratton, 'A New Switching and Transmission Plan for the Inland Trunk Network', *Post Office Electrical Engineers Journal* **53**, July 1960, pp. 75-9.
4. H. Williams, 'Overall Survey of Transmission-Performance Planning', *Proc. I.E.E.* **111**, 4, April 1964, pp. 727-43.
5. F. T. Andrews and R. W. Hatch, 'National Telephone Network Transmission Planning in the American Telephone and Telegraph Company', *Trans. I.E.E.E.* **COM-19**, June 1971, pp. 302-315.
6. H. R. Huntley, 'Transmission Design of Inter-Toll Telephone Trunks', *B.S.T.J.* **32**, September 1953, pp. 1019-36.
7. *C.C.I.T.T. White Book V*, Recommendation P42.
8. W. J. G. Mellos, 'Telephone Performance Measurement', *G.E.C. Journal* **34**, 1967, pp. 131-7.

9. J. Swaffield and R. H. de Wardt, 'A Reference Telephone System for Articulation Tests', *Post Office Electrical Engineers Journal* **43**, April 1950, pp. 1-7.
10. D. F. Hoth, 'Room Noise Spectra at Subscribers' Telephone Locations', *Journal of Acoustical Society of America* **12**, 1941, p. 499.
11. *C.C.I.T.T. White Book III*, Apendix to Section 1.
12. S. Munday, 'New International Switching and Transmission Plan Recommended by the C.C.I.T.T. for Public Telephony', *Proc. I.E.E.* **114**, 5, May 1967, pp. 619-27.
13. *C.C.I.T.T. White Book III*, Recommendation G111A.
14. *C.C.I.T.T. White Book III*, Recommendation G122.
15. R. L. Richards, 'Transmission Performance Assessment for Telephone Network Planning', *Proc. I.E.E.* **111**, 5, May 1964, pp. 931-40.
16. D. A. Lewinski, 'A New Objective for Message Circuit Noise', *B.S.T.J.* **63**, 2, March 1964, pp. 719-40.
17. *C.C.I.T.T. White Book III*, Recommendation G103.
18. *C.C.I.T.T. White Book III*, Recommendation G143.
19. *C.C.I.T.T. White Book III*, Recommendations G125, G152, G153 and G222.

Chapter 4

F.d.m. systems

4.1 Introduction

There is, in principle, no upper limit to the frequency which any smooth line can carry. The limitations come from the increase with frequency of attenuation and of vulnerability to pick-up of interfering signals. Hence the use of frequency division multiplexing techniques can increase the utilisation of transmission lines and reduce the number of amplifiers needed. The simplest form of f.d.m. is on open-wire routes where three to twelve channels may be carried, depending upon the relative cost of multiplexing equipment versus line costs. This has a wide range of applications in rural districts, particularly abroad, and nowadays the cost of terminal equipment has reduced to such an extent that it is becoming economic to use d.s.b. for short haul operations (less than 10 miles or 16 km) for local distribution[1].

The basic principles are well known and telegraph multiplexing was performed as early as 1918 using Campbell filters designed on the basis of artificial line theory. These were only constant-*k* sections but it was not the filters which were the limiting factor on multiplexing but repeater design. The repeaters used have to have very linear characteristics if the intermodulation of the different signals is to be avoided. Hence for a given degree of non-linearity, there is a limit to the number of channels that may be transmitted for a given level of intermodulation noise in any one channel. It was for this application that feedback was invented, gain stability was a secondary reason. In an f.d.m. system filters are needed at the receiver to separate the different signals, and if s.s.b. modulation is used then a filter at the sending end is required also to eliminate the unwanted sideband. Typically these filters must reject the unwanted signals by more than 60 dB and introduce only a fraction of a dB distortion into the pass-band.

The complexity and cost of an *LC* filter design depends upon its relative bandwidth, and since the absolute bandwidth required is fixed (about 3 kHz) then the lower the centre frequency, the simpler the filter. In the 1930s, *LC* filters could be made which worked at lower frequencies and three or four channels could be assembled. However, the development of piezo-electric filters before the war enabled efficient side-band selection to be made directly. The frequency range 60-108 kHz was chosen since this was best suited to the crystal filters—lower frequency crystals were large and expensive and difficult to obtain whereas higher frequency crystals were available but were difficult to make and adjust.

The 12-channel group thus became the basic building block of nearly all f.d.m. systems. If further channels are required, they may be provided by combining one or more groups into a supergroup and so on. This hierarchy of construction has the advantage of modularity of equipment and a limited number of different filter types. The major disadvantage is that it increases the number of frequency translations that a signal has to undergo, each of which introduces noise and distortion. The full multiplication table is[2].

Up to 24 telegraph sub-channels	= 1 channel	
12 channels	= 1 group	(12 channels = 48 kHz)
5 groups	= 1 supergroup	(60 channels = 240 kHz)
5 supergroup	= 1 mastergroup	(300 channels = 1·2 MHz)
3 mastergroups	= 1 supermastergroup	(900 channels = 3·6 MHz)

There are some variations: in North America the mastergroup consists of 10 supergroups and in Europe there is some use of a 15-supergroup assembly to make a supermastergroup. A practical system will carry a number of supermastergroups. Currently the largest practical system carries 12 supermastergroups (10 800 channels) and occupies a bandwidth of 60 MHz.

On some submarine cable systems a 3 kHz channelling system has been developed to give a 16-channel group. This reduction in required bandwidth is achieved by a slight reduction of the upper audio frequency and filtering of the audio frequency prior and post modulation[3].

Synchronisation of carrier frequencies

Since s.s.b. is used for multiplexing then it is necessary to have locally generated carrier signals. Each stage of modulation is equivalent to a frequency shift and can introduce a frequency error in the final received signal. For speech use a shift frequency of up to ±20 Hz is barely noticeable, but if the system is to be used for f.m. telegraphy or data, then tighter specifications are necessary. The CCITT recommend a maximum shift of ±2 Hz[4]. This implies an accuracy of up to one part in 10^8 for the highest modulation frequencies when allowance is made for the number of possible stages of frequency translation.

There are several ways in which a large system may be synchronised. The simplest is to use a pilot signal as a frequency reference at some central station and use this to drive the carrier generation equipment at all the other stations. This, of course, suffers from the security point of view in that the whole system fails if the main station or its links fail. To overcome this problem each station has its own carrier generator with some form of motor-driven frequency adjustment. The local frequency is compared with the frequency reference pilot (usually 60 kHz) and the difference is used to correct the local signal. If the received pilot fails for some reason, the system will continue to operate but will drift out of synchronism with time.

Nowadays the accuracy and stability of local oscillators is such that there is no need for continuous comparison and the oscillators may be set and only checked at regular intervals against the frequency standard.

4.2 Lines for h.f. transmission

Open-wire systems have been used for h.f. transmission and the main limitation is cross-talk and pick-up. If the wires are transposed at regular intervals then they can be used up to 150 kHz or even 300 kHz. A secondary disadvantage is the variation of attenuation with temperature and other weather conditions. Buried cables have a much larger attenuation per unit distance because of the higher capacitance per unit distance. This lowers Z_0 and as may be seen from Chapter 2 (page 36) if G is neglected then $\alpha = R/2Z_0$. At these higher frequencies of operation, R varies with $\sqrt{\omega}$ due to the skin effects. The velocity of propagation in cables is about 60 per cent lower than that in open-wire where it approaches the velocity of light. However, the majority of carrier systems use buried cables where the temperature variations are much less than those above ground (e.g. a cable 50 cm below the surface will suffer only 20 per cent of the open-air temperature variation) and this yields a more stable characteristic.

Up to 120 kHz it is possible to use twisted pair for carrier transmission. These are usually grouped together in a star-quad form in order to provide a high-grade phantom circuit. Modern cables consist of about 12 star-quads of 1·3 mm diameter (40 lb/mile) conductors and each quad has a different twist length so as to reduce mutual interference. A further reduction of this interference may be obtained by fitting capacitor networks at the receiving end of each cable section to balance out the effects of mutual interference due to capacitance and inductor unbalance. Typical attentuation figures are shown in Figure 4.1(a).

For higher frequency working the pick-up problem for twisted pairs becomes too great and it is necessary to go to a coaxial cable. This is an unbalanced system and so it cannot be used for lower frequencies (< 60 kHz) because of pick-up. For higher frequencies the outer casing acts as a shield since, due to the skin effect the carrier, current is confined to the inner wall, whereas any outside interference will cause currents to flow in the outside of the wall. The attenuation of the coaxial cable is less than that of a twisted pair made from the equivalent amount of copper.

The attenuation of the cable depends upon its dimensions. If the inner conductor radius is a and the radius of the inner surface of the outer conductor is b then the a.c. resistance per unit length is given by[5]

$$R_{ac} = \frac{1}{2\pi}\left(\frac{\omega\rho\mu}{2}\right)^{1/2}\left|\frac{1}{a}+\frac{1}{b}\right|$$

where ρ is the resistivity and μ is the permeability of the conductor.

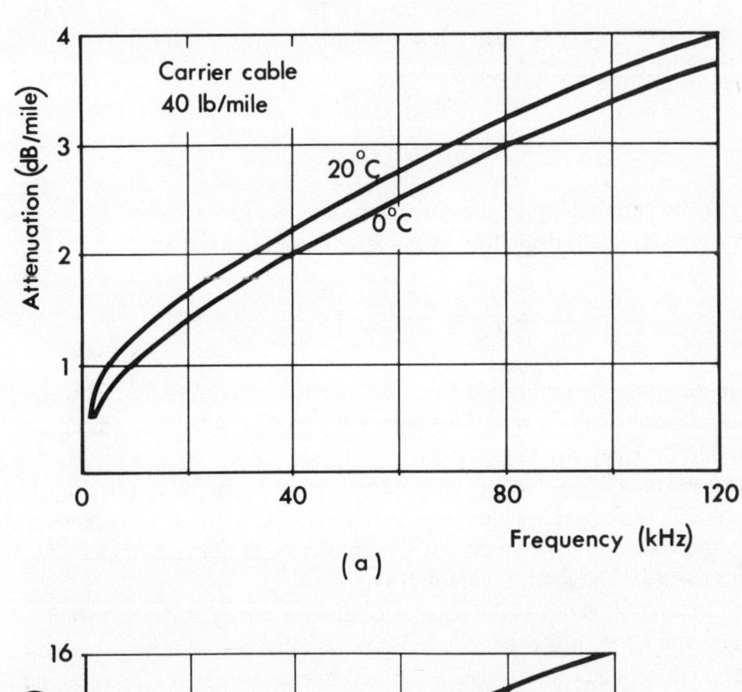

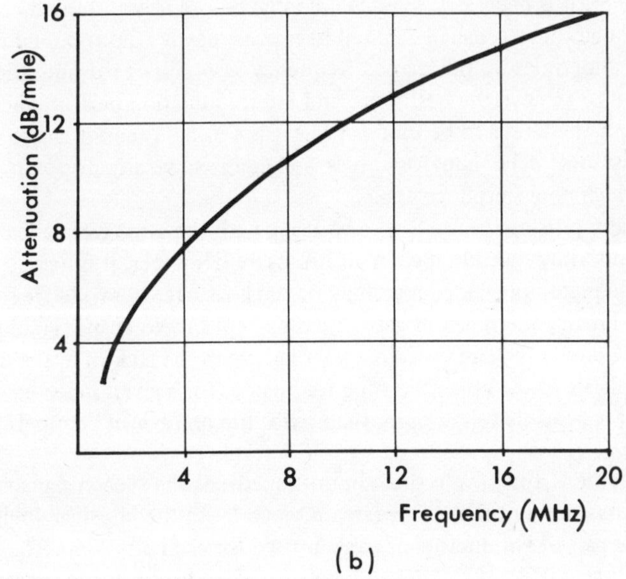

Figure 4.1 Attenuation of h.f. Cables
(a) carrier cable, (b) coaxial cable (9·5 mm)

At the frequencies of interest Z_0 is given by $(L/C)^{1/2}$ and for the coaxial line this is given by

$$Z_0 = \frac{1}{2\pi}\left(\frac{\mu}{\kappa}\right)^{1/2} \ln \frac{b}{a}$$

where κ is the permittivity of the dielectric.

G in a coaxial line is negligible so the attenuation is given by

$$\alpha = \frac{R}{2Z_0} = \frac{1}{2b}\left(\frac{\omega\rho\kappa}{2}\right)^{1/2} \frac{1 + b/a}{\ln (b/a)}$$

This equation may be minimised for a value of the ratio $b/a = 3{\cdot}6$. However, this minimum is very shallow, variations between 3·0 and 4·4 will only alter the attenuation by 1 per cent[5].

With a value of $b/a = 3{\cdot}6$ the characteristic impedance for an (effectively) all-air cable is 75 Ω over a wide range of frequencies. With other dielectrics the attenuation constant is increased but the physical stability is increased. Typical characteristics are shown in Figure 4.1(b).

The attenuation varies as the square root of frequency above a certain frequency and hence it is necessary to equalise a line before it is amplified. In practice the equalisation is usually shared between a passive input circuit and the amplifier, or sometimes by the shaped frequency response of the amplifier itself. This is done to achieve the best noise and intermodulation performance of the combination. If the line is to be used for television or data then it is also necessary to apply some delay equalisation as well and remove any impedance discontinuities which could produce echos.

It is theoretically possible to use 2-wire working with the same frequencies in both directions and using hybrid repeaters if necessary. However, it is very difficult to get adequate impedance matching of the lines over the wide frequency range, and hence the repeater gains must be kept low to ensure stability. Since the attenuations of the cables follow a $\sqrt{f}$ law, then the lack of available amplification implies a lower upper working frequency (for a given repeater spacing). Hence it is possible to get more than twice the number of channels in a 4-wire system.

The 4-wire may be virtual or physical. In both methods the speech signals are divided by audio hybrids and thereafter remain separate. In the physical 4-wire system, a separate pair of conductors or cable is used for each direction. The equivalent 4-wire uses different frequency bands for each direction of travel and, at each repeater, high-pass and low-pass filters are used to separate the two directions of travel.

The equivalent 4-wire has the disadvantages that a guard band has to be left in the middle and also that the filter impedances present a bad impedance to the line and hence produce echos. This technique is only used for very low capacity systems or for long submarine systems.

The upper frequency limit of a cable system is governed by the repeater spacing and their gains. For a given cable the signal must be amplified before the highest frequency is too contaminated by noise or cross talk. The limitations to the gain of a repeater are discussed later.

Cross-talk

Cross talk can be intelligible or may be unintelligible if it comes from a channel with frequency inversion. In the latter case the interference retains the syllabic pattern of speech and is therefore more annoying than the white noise of the same mean power. In either case it is due to three main causes[6] (see Figure 4.2).

(a) *Near-end cross-talk* (NEXT) This is due to pick-up from cables in opposite directions; the simplest solution is physically to separate the two directions or else to use different frequency bands for the two directions of travel.

(b) *Far-end cross-talk* (FEXT) This is due to the output of one amplifier to the input of another for signals travelling in the same direction due to the proximity of the cables. If there are only two systems, then one technique for reducing this cross-talk is to reverse the phase of one of the systems at regular intervals and this will produce a partial cancelling effect. Another technique would be to use opposite sidebands in each system which will not reduce the cross-talk but will render it unintelligible and hence make it less troublesome.

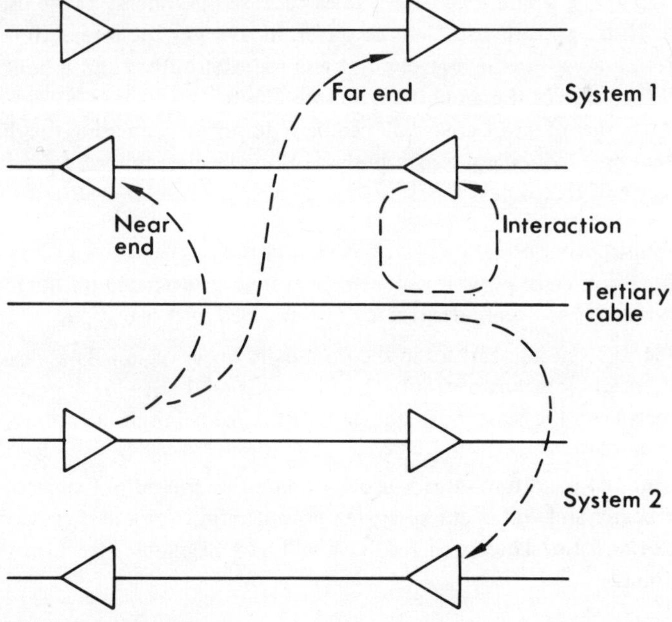

Figure 4.2 Cross-talk paths

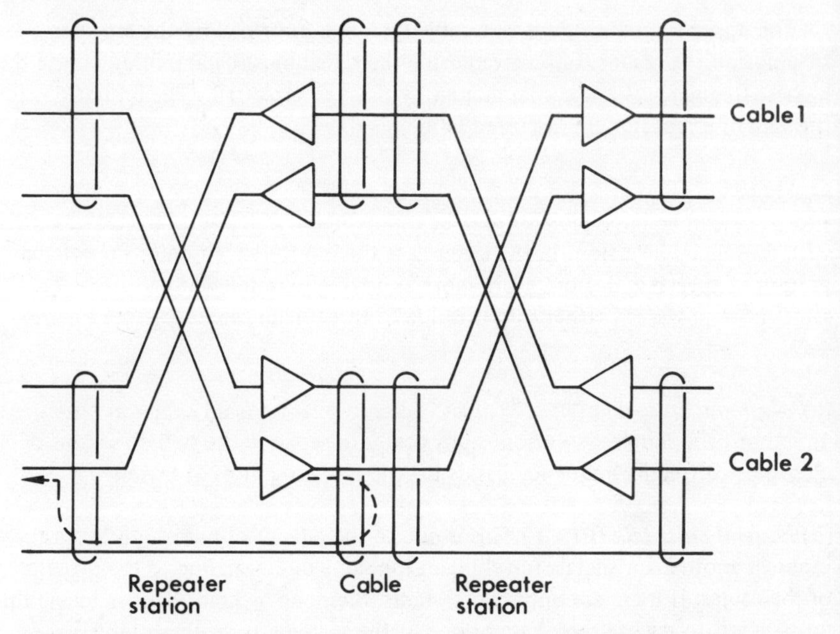

Figure 4.3 Technique for reducing near-end and interaction cross-talk

(c) *Interaction cross-talk* This is caused by coupling via a third conductor which might be a shield or a voice frequency pair in the same cable. This is reducible if there are separate GO and RETURN cables because the cables may be used alternatively for opposite directions of travel. In this way the interaction path is made to terminate at the high-level point at a repeater output and is hence less serious by the value of the amplifier gain (see Figure 4.3).

In all cases the effect of cross-talk can be reduced by decreasing the distance between sections, reducing the maximum working frequency, or by producing a better balance of the cables.

Effect of repeater imperfections–intermodulation

An additional source of noise in a multiplex system is produced by the repeater non-linearities. These non-linearities may be grouped into two types:

(i) the slight non-linearities in the nominally linear region. These can be reduced by means of feedback but increasing the feedback necessitates increasing the bandwidth and gain of the main amplifier to prevent instability.

(ii) overload condition–this is usually limited by the output stage of the repeater and will occur at a given power output for a particular valve or transistor. The use of feedback will have negligible effect on this point.

In general the overload point is an arbitrarily defined one. For instance, one of the definitions adopted by the CCITT is 'the overload level of an amplifier is that value of absolute power (in dBm) at the output, at which the absolute power of the third harmonic increases by 20 dB when the input signal increases by 1 dB'. In a repeater with a large amount of negative feedback (> 40 dB) then once the overload point has been reached, the performance deteriorates rapidly and hence this point is easily distinguished.

The effect of slight non-linearities will produce intermodulation distortion which will show as noise in a telephone channel. If the repeater transfer function can be expressed as a power series below overload,

$$e_0 = a_1 e_i + a_2 e_i^2 + a_3 e_i^3 + \cdots$$

then the higher order terms represent the distortion due to the non-linear properties of the devices in the repeater.

Consider the effect of the repeater on a signal containing two components,

then $\quad e_i = A \cos \alpha t + B \cos \beta t$

$$\begin{aligned} e_0 = {} & \tfrac{1}{2}a_2 (A^2 + B^2) + a_1 [A \cos \alpha t + B \cos \beta t] \\ & + a_2 [\tfrac{1}{2}A^2 \cos 2\alpha t + \tfrac{1}{2}B^2 \cos 2\beta t + AB\{\cos(\alpha + \beta)t + \cos(\alpha - \beta)t\}] \\ & + a_3 [\tfrac{3}{4}A(A^2 + 2B^2) \cos \alpha t + \tfrac{3}{4}B(B^2 + 2A^2) \cos \beta t \\ & \qquad + \tfrac{1}{4}A^3 \cos 3\alpha t + \tfrac{1}{4}B^3 \cos 3\beta t \\ & \qquad + \tfrac{3}{4}A^2 B\{\cos(2\alpha + \beta)t + \cos(2\alpha - \beta)t\} \\ & \qquad + \tfrac{3}{4}B^2 A\{\cos(2\beta + \alpha)t + \cos(2\beta - \alpha)t\}] \\ & + \cdots \end{aligned}$$

and the effect is to introduce components at new frequencies and if these new frequencies are also within the transmitted band then they will appear as noise. The wider the transmitted band the greater the noise generated in any one particular channel, and this is a fundamental limitation on the channel capacity of a system.

If both fundamentals are increased by 1 dB then the magnitude of all the second order factors will be increased by 2 dB (because the coefficients all have terms like A^2, AB etc.) and all third order factors will increase by 3 dB etc. Hence the relative magnitude of the harmonics depend upon the absolute level of the fundamentals.

In fact it is very difficult to measure the coefficients of the power series directly; what can be measured is the distortion produced by one and two-tone signals. It is then possible to define modulation coefficients M_2, M_3 etc. as the ratio (in dB) of the second- and third-order distortion products to the fundamental at a given output level (e.g. 1 mW). Since the distortion is usually prodominantly produced by the output stage, then these coefficients are independent of the gain of the repeater[7].

If these are the coefficients for the basic amplifier then those for a system with feedback M_{2R}, M_{3R} will be reduced. The second order term will in fact be reduced by the feedback factor $F = 20 \log | 1 - \mu\beta |$ dB, where μ and β have their usual significance. However, the third order terms will not have such a reduction because the second-order distortion which does occur combines with the fundamental to produce further third-order terms.

It is possible to compute the effects of the non-linearities on the noise in one channel[8, 9, 10], but it is usually sufficient to measure it on a representative repeater with a suitable test signal as will be described later.

When there are several repeaters in tandem then each will contribute its own intermodulation products and these will add overall. In order to find how they add, consider the situation shown in Figure 4.4(a), where it is assumed that the

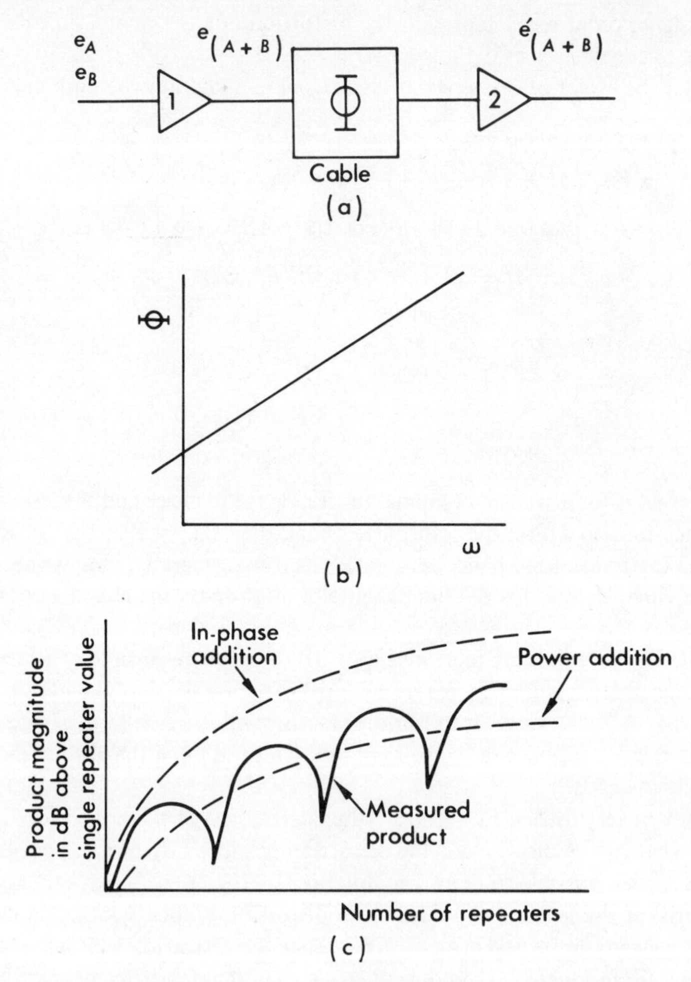

Figure 4.4 Addition of intermodulation products
(a) basic situation, (b) phase characteristic of cable, (c) second-order modulation product variation

cable and equaliser have a linear phase characteristic as shown in Figure 4.4(b), i.e. the phase shift is given by

$$\Phi = a\omega + b$$

over the working frequency band.

Consider first the second-order products produced from the fundamentals,

$$e_A = \cos \alpha t$$

$$e_B = \cos \beta t$$

The second order product at the first amplifier will be of the form

$$e_{A+B} = k \cos(\alpha + \beta)t$$

At the output of the second amplifier this component will be

$$e'_{A+B} = k \cos[(\alpha + \beta)t + a(\alpha + \beta) + b]$$

At the input to the second amplifier the fundamentals will be

$$e'_A = \cos[\alpha t + a\alpha + b]$$

$$e'_B = \cos[\beta t + a\beta + b]$$

So the second-order product produced by the second amplifier will be

$$e''_{A+B} = k \cos[(\alpha + \beta)t + a(\alpha + \beta) + 2b]$$

i.e. this will add in phase with the product from the first amplifier if $b = 0$ or $2\pi n$. Since in general the length of cables between repeaters have slight variations in length, the effect of addition will be random and a good approximation is found by assuming that the second-order components on a multi-section cable add in a power basis (see Figure 4.3(c)).

Performing a similar exercise for third-order products with another fundamental,

$$e_C = \cos \gamma t$$

then the third-order products will be of the form typically

$$e_{A+B-C} = k \cos(\alpha + \beta - \gamma)t$$

and at the output of the second amplifier this component will be

$$e'_{A+B-C} = k \cos[(\alpha + \beta - \gamma)t + a(\alpha + \beta - \gamma) + b]$$

The third-order product produced by the second amplifier will be of exactly the same form and hence these third-order products will tend to add in phase and hence on a voltage basis rather than power. This will not be true for intermodulation terms of the form e_{A+B+C}, e_{2A+B} or e_{3A}, but these provide a small proportion of the third-order products.

In an equivalent 4-wire system, the presence of the band-separation filters produces sufficient phase distortion to make power addition a reasonable approximation for all terms. The design of suitable feedback amplifiers and their equalisation is a subject on its own [11,12].

In order to estimate the requirements on repeater linearities, it is first necessary to obtain some further details on the nature of the multiplex signal.

4.3 Characteristics of a multiplex signal

There are two main parameters that must be estimated for a multiplex signal:

(a) average power in the busy hour—to enable the intermodulation noise to be computed.

(b) instantaneous peak value—to ensure that it is below the overload point. For an indeterminate signal the absolute peak value would be very high, so some alternative such as that value which is never exceeded more than 0·01 per cent of the time is used. The actual percentage will depend upon the characteristics of the amplifier used and must be found by experiment.

These figures can be obtained only from extensive experimental observation. The first figure required is the average power of an individual talker measured at a suitable zero reference point.

It is convenient to split this mean power into two components:

(i) mean power whilst active
(ii) activity ratio

Conversational speech on a telephone is found to consist of short utterances averaging about 1 s in duration and that speaking and listening roles alternate some 15 times a minute. One objective method of deciding when speech is present is to count silence as those periods of at least 350 ms for which the short-term mean power (averaged over 10-20 ms) remains 15 dB below the mean power whilst active [13]. If the mean power whilst active is z mW then the volume is defined as $y = 10 \log_{10} z$ dBm, i.e. a fixed value for a given speaker and connection. It is necessary to define where the measurement should take place. This is usually chosen to be the 2-wire entry point at the first exchange in the hierarchy, i.e. the group switching centre in the United Kingdom or toll centre in North America. The reason for this choice is that the transmittion between local exchange and the group switching centre is usually on a physical 2- or 4-wire basis (or more recently by p.c.m.), but from the secondary centre the long distance network is almost universally by f.d.m. systems. It is found from observation of the distribution of the volumes measured at these points has a normal probability distribution (i.e. a log-normal distribution of powers). This takes into account the variation between actual speaker volumes and the spread of reference equivalent values to the measuring point.

The mean value of this distribution, $\bar{y}$, in the United Kingdom system is around -14 dBm0, i.e. this is the volume of the median subscriber. The standard deviation of the distribution is approximately $\sigma_y = 5$ dB. These figures may be converted to an average mean power per channel provided the activity ratio is known. For an inland system the activity ratio is found to be around 0·35 whilst the circuit is in use. Since a circuit even in the busiest hour is only occupied about 70-80 per cent of the time then the overall activity ratio drops to around 0·25.

Since the distribution of powers is log-normal then the average power will not be the same as the average volume. It may be shown[14] that the average power is related to the average volume by

$$P_{av} = \bar{y} + 0{\cdot}115\sigma_y^2 \text{ dBm0}$$

$$= -11{\cdot}1 \text{ dBm0}$$

For an activity ratio of 0·25 the long-term busy-hour average power reduces to $-17{\cdot}1$ dBm0 for speech. Signalling tones average 10 μW long term, i.e. -20 dBm0 so the total mean power is $-15{\cdot}3$ dBm0.

The figures found by other administrations[15,16] differ slightly from these since they have different microphones and distribution of reference equivalents, and they also are likely to have different telephoning habits.

For the purpose of design, the CCITT have adopted a *conventional* value of the mean absolute power (at zero relative level) of the speech plus signalling tones etc. transmitted over a telephone channel in one direction of transmission during the busiest hour[17]. This is chosen to be 31·6 μW0 (= -15 dBm0) based on a mean power of about 22 μW0 for the speech and 10 μW0 for the signalling tones etc. Changes in the network will change the values actually present at the entry to the trunk network. Also the advent of more non-telephonic services on the speech network such as multi-channel telegraphy and data would change this loading. What in fact happens nowadays is that any addition to the network or any new service is designed such that the long-term mean power it produces at the trunk network entry is not greater than -15 dBm0. Hence this figure has now become a design figure rather than a measurement.

The mean power of an N-channel multiplex signal is thus

$$P_N = -15 + 10 \log N \text{ dBm0}$$

Peak power

The other factor is that of peak signal. The original measurements of the relationship between the mean and the peak power of a telephone speech signal were performed by Holbrook and Dixon in 1938[18]. These measurements have been repeated with time, and attempts have been made to find theoretical models for a combination of many speakers[19]. As a result of these measurements it is possible to construct a table giving the power of an equivalent sine wave signal which has the same peak power as N talkers[20] (Table 4.1). An alternative way of presenting these results is to give a graph of the peak to mean

TABLE 4.1

Power of sine wave having the same peak as a multiplex signal at zero relative level, assuming mean power is −15 dBm0, $\sigma = 5{\cdot}8$ dB

Channel	12	24	36	48	60	120	300	600	960	1800	2700
dBm0	19	19·5	20	20·5	20·8	21·2	23	25	27	30	32

ratio, or *multi-channel load factor* as it is usually called (Figure 4.5). Notice that for a large number of channels it tends to a value of 11 dB.

If, due to a fault or mis-operation, the signal level on any one group or supergroup is allowed to exceed the mean or the peak power, then overloading will occur and this could lead to the loss of a large number of circuits. To prevent such overloading, modern equipment has group and supergroup limiters applied to the signal prior to their assembly into the higher multiplex. These limiters operate by introducing a loss to compensate for any increase in peak power or the mean power. In practice recognition times of the order of 1 ms are used with a restoration of about 5 ms.

Load simulation

It is frequently desirable to have a test signal which may be used in place of a multiplex signal, in particular for measuring intermodulation noise. The amplitude distribution of the sum of a large number of telephone speakers are shown

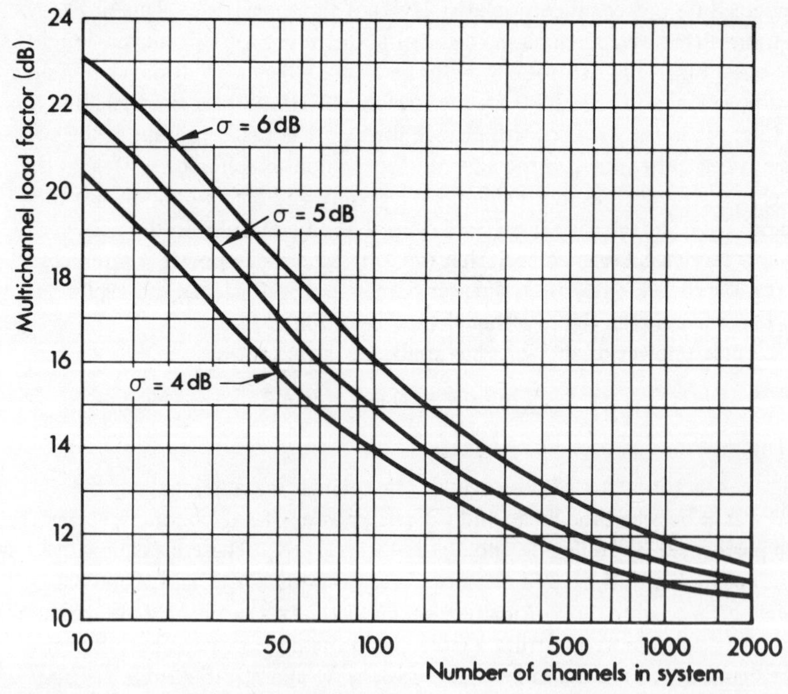

Figure 4.5 Multichannel load factor

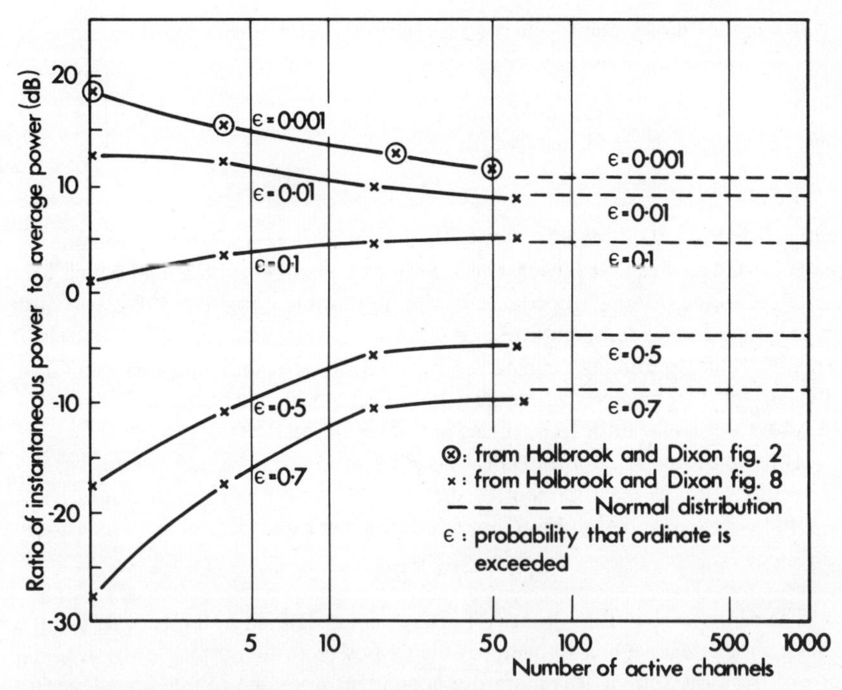

Figure 4.6 Amplitude distribution of a large number of telephone speakers (taken from R. G. Medhurst, 'Distortion in Microwave trunk-radio systems', *G.E.C. Journal*, **34**, 1967, pp. 75-83)

in Figure 4.6. As may be seen when the number of talkers exceeds about sixty these figures approximate those of a normal distribution; hence in these cases white noise is a good approximation to this signal and the relevant power of equivalent noise signal is

$$L_N = -15 + 10 \log N \text{ dBm0} \qquad \text{for } N > 240$$

i.e. the same as the mean power. Below $N = 240$ the statistics differ from pure random noise but a good approximation is given by white noise at a power level of

$$L_N = -1 + 4 \log N \text{ dBm0} \qquad \text{for } 60 < N < 240$$

The intermodulation behaviour can now be measured by using a white noise signal of the suitable level and bandwidth. If a frequency gap one channel wide in the noise is produced, then the effect of intermodulation will be to produce noise in this channel and this can be measured at the end of the system.

4.4 Noise in cable systems

There are two irreducible major causes of random noise in any electronic system:

(a) thermal noise, due to the thermal motion of electrons in a conductor,

(b) shot noise, due to the discrete nature of electrons when they arrive at an anode or collector.

Thermal noise depends only upon the temperature of the device. The available power in a 1 Hz bandwidth is given by

$$p = kT$$

where k is Boltzman's constant and T is the absolute temperature. Available power is defined as that power which would be absorbed into a load which was matched to the internal impedance of the noise source. For $T = 290$ K (usually called the standard temperature), $p = 4 \times 10^{-21}$ watts/Hz, i.e. -174 dBm/Hz. This relationship implies a constant power density spectrum and this is in fact the case until frequencies where quantum-mechanical effects occur. However, this does not occur until frequencies in the region of 1000 GHz ($\text{G} \equiv 10^9$).

The use of this relationship allows the definition of *noise temperature* as a convenient unit of measurement. If the available power in a bandwidth of B Hz is p then the equivalent noise temperature is given by

$$T = p/kB$$

The noise temperature of a two-port system is defined in terms of its available gain, g (this is the ratio of the available power at the output to the available power from the source; it is therefore dependent upon the source impedance). If the two-port is connected to a noise source with a noise temperature of T then the output noise p_o will consist of the amplified input noise plus the internally generated noise p_{int}, i.e.

$$p_o = gkTB + p_{int}$$

If the effect of the internally generated noise is replaced by adding an additional noise source at the input of an equivalent noiseless two-port then this additional noise source will have a noise temperature of

$$T_{\text{int}} = P_{\text{int}}/gkB$$

and hence

$$p_o = gk(T + T_{\text{int}})B$$

so the effective input noise temperature will be $T + T_{\text{int}}$. Note that the noise temperature is not the same as the physical temperature and that it can be a function of frequency.

An alternative unit is that of *noise factor* of a two-port system. This is defined as the ratio of the total output of noise power in a specified bandwidth when the input has a noise temperature of 290 K to that portion of the output noise power due to the input source above, i.e.

$$n^F = \frac{p_o}{gkT_oB} \qquad \text{when } T_o = 290 \text{ K}$$

This usually is expressed in dB, $N_F = 10 \log n_F$. It is easy to show that noise factor and noise temperature are related by

$$T_{\text{int}} = T_o\,(n_F - 1)$$

and

$$n_F = 1 + \frac{T_{\text{int}}}{T_o}$$

The noise factor concept is usually more convenient when a simple system with an input source whose noise temperature is near 290 K but noise temperature is more useful when the overall performance of a complex system such as a satellite link is considered (see Chapter 8).

Noise in a repeater section

A repeater section consisting of a length of cable, its equaliser and repeater can be considered as a simple system. The cable will normally be around 290 K and hence the output noise power from a repeater will be given by

$$p_o = n_F gkTB \text{ watts}$$

where n_F is the noise factor of the repeater and g is the available gain of the equaliser and repeater. Since a repeater is matched to the cable to prevent reflections, then the actual gain is equal to the available gain. In dB this gives

$$P_o = N_F + G - 174 + 10 \log B \text{ dBm}$$

where G is the repeater + equaliser in dB. For $B = 4$ kHz then $10 \log B \triangleq 36$ dB so that the output noise power of a repeater into a 4 kHz band is given by

$$P_o = N_F + G - 138 \text{ dBm}$$

The repeater will normally be used to make up the losses of the cable so that the complete repeater section will have a nominal 0 dB loss at all frequencies. The effect of deviations from this will be considered later. If a system consists of a number of repeater sections then the noise from each section of n identical sections will be np_o, i.e. if this power is P_R dBm then

$$P_R = N_F + G - 138 + 10 \log n \text{ dBm}$$

If the relative output level of the repeater is C dBr then the noise referred to the zero reference point will be

$$P_{R0} = P_R - C \text{ dBm0}$$

EXAMPLE 1

For a 2000 km system consisting of 100 repeater sections, if the repeater gain at the maximum frequency is 50 dB and the noise factor at this frequency is 10 dB, then the total noise from the system is

$$\begin{aligned} P_R &= 10 + 50 - 138 + 20 \text{ dBm} \\ &= -58 \text{ dBm} \end{aligned}$$

If the output level of the repeater is –5 dBr then the noise referred to the zero reference point will be –53 dBm0. Since the power in a 4 kHz band of white noise is reduced by 3·6 dB when it is psophometrically weighted then –53 dBm0 corresponds to –56·6 dBm0p, i.e. 2190 pW0p. This is equivalent to 1·09 pW/km for the 2000 km circuit.

Note that the effect of increasing the relative level at the repeater output is to decrease the effective noise contribution at the zero reference point.

EXAMPLE 2

Find the minimum number of repeaters for a 300 channel cable system of 1000 km length, whose top frequency attenuation is L = 1000 dB, designed for a weighted thermal noise contribution of 2000 pW per channel at the zero reference point. Assume the repeaters have a 6 dB noise figure and have an overload point of 20 dBm at their output.

The equivalent peak power of 300 channels is 23 dBm0, and hence the maximum output level of the repeaters must be –3 dBr.

$$2000 \text{ pW0p} = -57 \text{ dBm0p} = -53{\cdot}4 \text{ dBm0}$$

Since the system must operate at –3 dBr then the total noise must be –56·4 dBm. If there are n repeaters then their individual gain must be $G = 1000/n$ dB so the total number of repeaters is given by the minimum integer satisfying the equation

$$N_F + G - 138 + 10 \log n < P_R$$

i.e.

$$\frac{1000}{n} + 10 \log n < 75{\cdot}6$$

This equation may be solved by trial and error or graphically as shown in Figure 4.7 giving $n = 16$.

Choice of f.d.m. system level

The most important parameter in a repeatered f.d.m. system is the relative level at which it operates. The signal-to-noise ratio is normally worst at the highest frequency channel, and for this reason some degree of pre-emphasis is usually given to the signal. However, as far as the transmission system is concerned, this has a flat frequency response over its working range and the noise behaviour may be investigated by considering the noise in the highest frequency channel.

It is clear from the previous section that increasing the system level will decrease the effect of thermal noise. However, from section 4.2 it may be seen that increasing system level will increase the intermodulation noise.

This may be seen from the following argument: assume that the repeater is loaded with a white noise signal of the appropriate power level, P_N dBm, and bandwidth for the required number of channels. Then for this input the noise powers due to second- and third-order intermodulation products are (say)

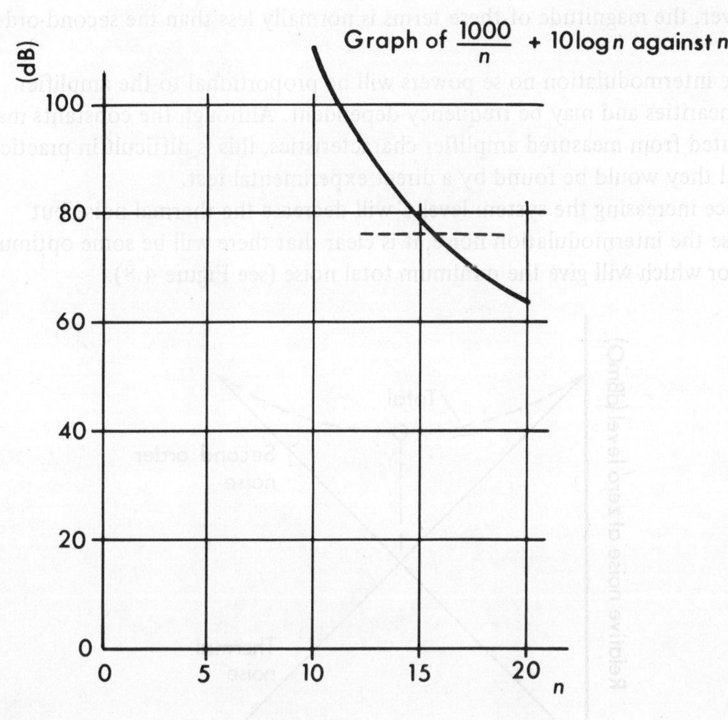

Figure 4.7 Graph of $\frac{1000}{n} + 10 \log n$ *v. n*

M_{2R0} dBm and M_{3R0} dBm respectively in some particular channel. If the working relative level of the repeater output is C dBr then the output power level will be $(P_N + C)$ dBm and, as explained in section 4.2, the intermodulation noise power will be changed to

$$M_{2R} = M_{2R0} + 2C \text{ dBm}$$

$$M_{3R} = M_{3R0} + 3C \text{ dBm}$$

However, for a relative level of C dBr the noise contribution referred to the zero reference point will be C dB less than this, i.e. $(M_{2R0} + C)$ dBm0 and $(M_{3R0} + 2C)$ dBm0.

If the noise power from the individual repeater sections add on a power basis (as it is found do the second-order terms in general) then the total noise due to second order intermodulation will be

$$M_{2R} = M_{2R0} + C + 10 \log n \text{ dBm0}$$

In general, the third-order terms are found to add on voltage basis. Hence for this case

$$M_{3R} = M_{3R0} + 2C + 20 \log n \text{ dBm0}$$

However, the magnitude of these terms is normally less than the second-order terms.

The intermodulation noise powers will be proportional to the amplifier non-linearities and may be frequency dependent. Although the constants may be computed from measured amplifier characteristics, this is difficult in practice. In general they would be found by a direct experimental test.

Since increasing the system level C will decrease the thermal noise but increase the intermodulation noise, it is clear that there will be some optimum level for which will give the minimum total noise (see Figure 4.8).

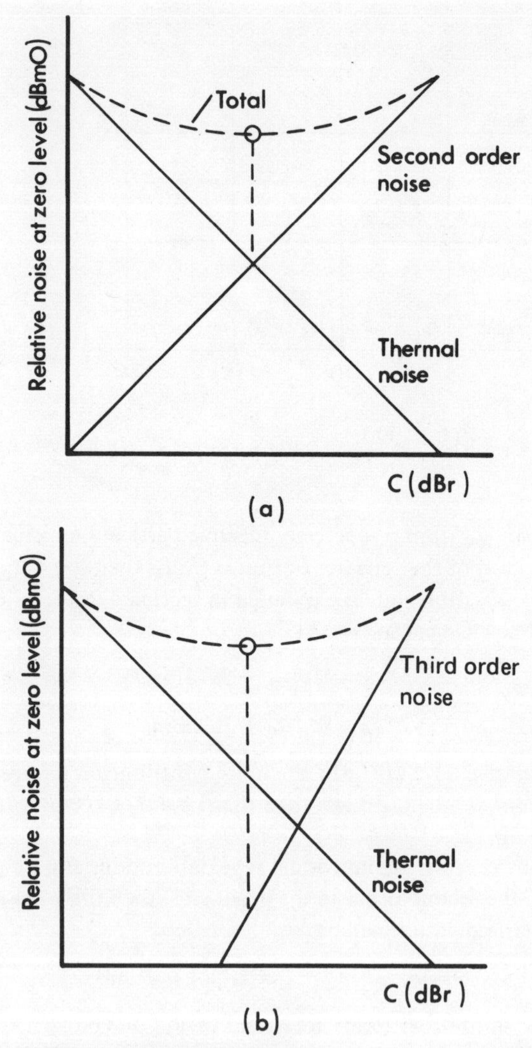

Figure 4.8 Optimum noise relationships
(a) second-order noise, (b) third-order noise

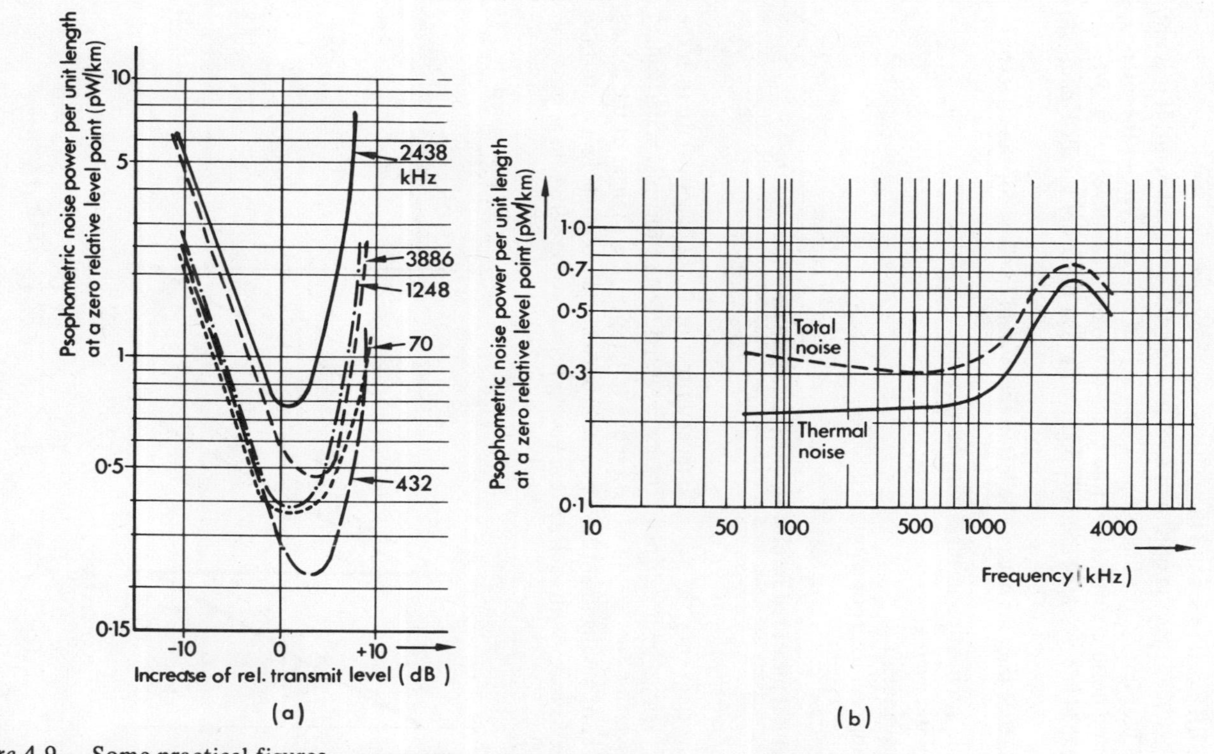

Figure 4.9 Some practical figures
(from K. H. Lehnich, 'Compatible 300 and 960-channel line equipment for small diameter coaxial cable', *Trans. I.E.E.E.* **COM–15**, 1, February 1967, pp. 108–109)

For the second-order noise, the minimum total may be shown to occur when the noise powers are equal and hence the total will be 3 dB above the thermal noise. For the case of third-order noise the minimum is found to occur when the thermal noise is 3 dB greater than the modulation noise, and the total noise is then 1·8 dB greater than the thermal.

This assumes that the optimum level is below the overload point. If it is not then the best level is the highest possible level. In modern systems the inter-modulation noise is sufficiently low to make the optimum level below overload. This is called the *overload limited* condition whilst the other is called *modulation limited.*

In practice margins have to be left in the system to allow for the effects of

–cable attenuation changes with temperature,
–repeater ageing,
–misalignment,
–changes in usage of equipment,
etc.

Some typical measured results on one system are shown in Figure 4.9.

Effect of misalignment

If at some frequency the gain of a repeater section is not exactly 0 dB then the system is said to be misaligned. There are several reasons for this, in particular, inexact equalisation or gain variation due to temperature changes of the cable. The effect of misalignment may be seen from a simple example of a uniformly misaligned system with each section having the same non-zero gain as shown in

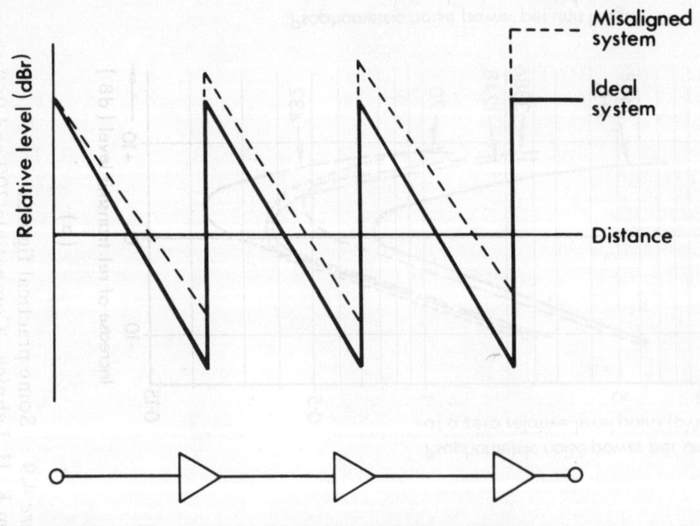

Figure 4.10 The effect of misalignment (positive)

Figure 4.10. At each repeater the relative level will increase, and had the system been optimum before the misalignment then the total noise contribution will increase with the change in level. This effect can be quite considerable and puts a premium on keeping a system aligned by very exact equalisation and a number of variable gain repeaters to remove the effect of the temperature variation of the cable loss. For instance it may be shown[21] that if a system is uniformly misaligned by a total of 10 dB (i.e. total loss is 10 dB rather than 0 dB) then this will increase the total noise by nearly 8 dB. Obviously a practical case of random misalignment is more complex and usually involves a computer simulation to compute the effect.

4.5 Hypothetical reference circuit

In order to set design objectives for carrier systems, the CCITT have introduced the concept of 'hypothetical reference circuit'[22]. This is a similar concept to the hypothetical reference connection used in transmission planning, as described in the last chapter. They consist of a representative circuit of a specified length with a large but not maximum possible number of frequency translations. The reference circuit is divided into a number of homogeneous sections in which it is assumed that no frequency translations occur. Within these sections certain intermodulation and cross-talk signals can add on a voltage basis, but the effect of frequency translation and the random interconnection of channels at the ends of each section prevents this voltage addition from being continued for several sections.

Originally, the nominal length of this circuit was 2500 km and was meant to represent a long international connection; it was divided into 9 equal lengths called homogeneous sections (i.e. about 280 km or 172 miles long). Within the overall length, it is assumed that there are three pairs of channel modulation stages (i.e. it is switched at audio and at two intermediate points) and also six pairs of group and nine pairs of supergroup modulation stages. This is shown diagrammatically in Figure 4.11. Suitable economic allocations of the noise and frequency impairments can now be allocated to each of the components in this hypothetical circuit, and these can then form the basis of design objectives. If a system satisfying these requirements is now used for a part of an international connection, there will be a high probability that the system will perform satisfactorily.

Nowadays, with international circuits, round the world connections can lead to distances of the order of 25 000 km and this is used as the basis of a new h.r.c. In order to achieve satisfactory performance over such long distances, it is necessary to have transmission systems which contribute only 1 pW/km for the majority of their length and this frequently implies some selection of the circuits that are to be used for intercontinental extensions.

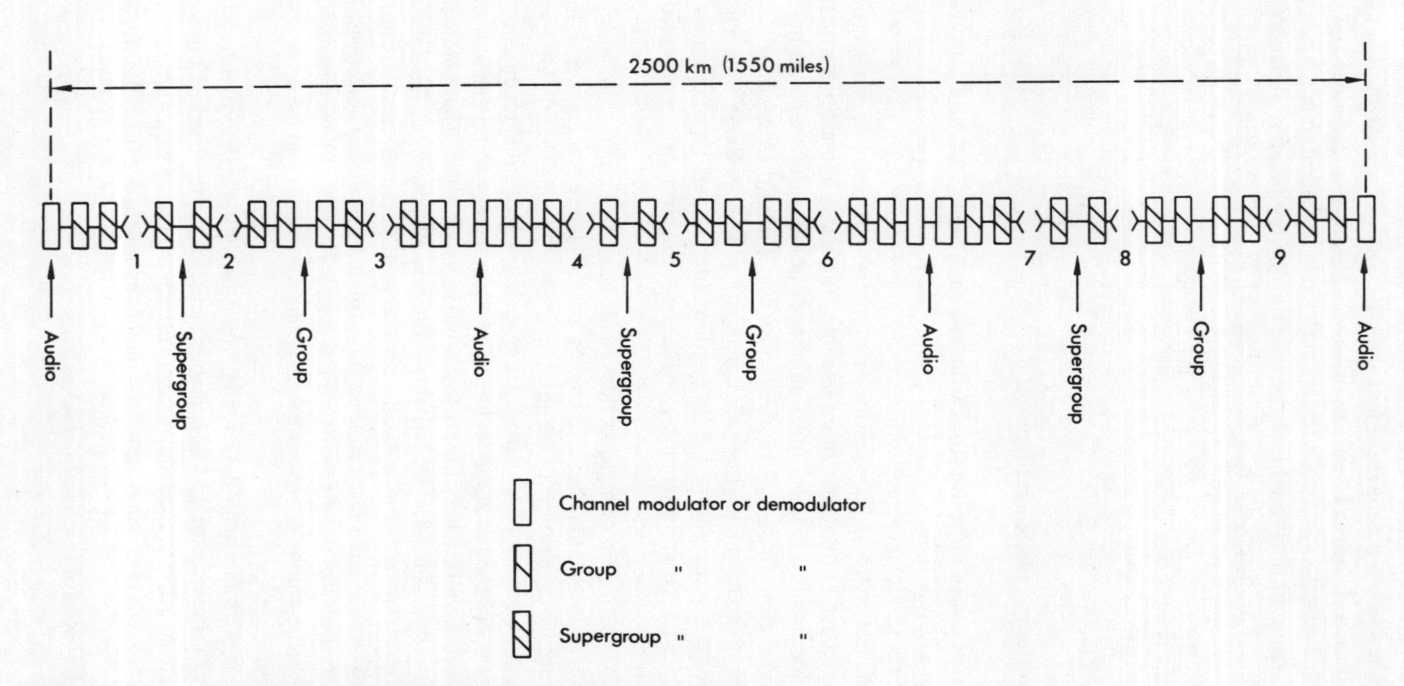

Figure 4.11 An example of a hypothetical reference connection

4.6 Practical aspects of f.d.m. systems

Power feeding

With frequent repeater spacing, it is necessary to reduce the number of places where a power source is needed. For this reason the power for the repeaters is normally fed at d.c. along the centre conductors of a pair of cables and the repeater power supply circuit is placed in series with cable. The supply is regulated to produce a constant current, so with the older type of repeater, which uses valves, voltages of the order of 100 V each are needed; this necessitates very high feeding voltages and so good insulation of the cable is necessary if many repeaters were to be driven in series. A further disadvantage of the high voltage is that as they are lethal then the whole system has to be turned off before any repairs could be done to any cable in a duct. Safety regulations in the United Kingdom provide for a key to open the manhole which is kept in the nearest power feeding repeater station, and removal of the key automatically removes the power. Hence, any repair implies a long out of service time.

The use of transistorised repeaters reduces the overall voltage requirements and it is possible to increase the number of repeaters and still have an intrinsically safe power supply.

Regulation

The loss of a cable will change with time due to temperature changes and to ageing effects. In general the change of loss will be frequency dependent. The calculations earlier in this chapter showed that loss stability of better than ±0·5 dB is needed for a complete link, and this implies that the loss/frequency characteristics of cable plus equaliser and amplifier must be constant to an order of magnitude better than this over the complete frequency range of the f.d.m. system.

This is achieved by the use of fixed frequency pilots of known amplitude being inserted in the multiplex signal. Selected repeaters may be arranged to have a pilot detection circuit which operates an automatic gain control to bring the pilot level back to its correct level. This will correct the majority of loss variations but the frequency dependent changes in loss require more sophisticated circuitry and uses two or three pilots covering the complete frequency range. The difference in received levels will then give an indication of the degree of frequency distortion and this information may be used to vary a suitable equaliser.

Ideally each repeater ought to be regulated, otherwise a misalignment will occur and increase the noise level. However, the cost of a regulated repeater is high and, more importantly, they have a much greater power requirement. Hence their use has to be minimised. For underground cables a proportion of the regulation may be performed by making the amplifier gain temperature-dependent, which is normally achieved by using a thermistor in the feedback network. This is much cheaper and uses less power than a pilot-regulated system and makes the necessity for pilot controlled regulation less frequent.

In addition to these automatic corrections it is usually necessary in a wide-band system to provide a manually operated equalisation at the terminal stations (or in a very long distance system at intermediate power feeding stations as well). This normally involves a complex remote control system for measurement and adjustment of the cable characteristics in order to permit the setting up from one station[23].

Submarine systems
Nearly all submarine systems are equivalent 4-wire, i.e. one direction of travel is in a low-band and the return direction is in the high-band. This permits the use of a common amplifier as shown in Figure 4.12. For very wide-band systems (above 5 MHz) this arrangement has proved troublesome, since an overload instability condition has been found whereby power transfer occurs between the high and low bands if the common amplifier is overloaded. At high bandwidths this condition is found to be self-sustaining under certain conditions, even after the original overload has been removed. Hence for these newer systems an amplifier per direction is needed.

The main requirement for submarine systems is, of course, extremely high reliability, since repair is so expensive. This has involved the development of special valves and transistors as well as other components to have a life in excess of twenty years. In the valve repeaters a measure of redundancy was introduced by using two amplifiers in parallel with a common feedback circuit. However, modern transistor amplifiers are sufficiently reliable to use only one amplifier in each repeater.

Since a cable may only be fed from either end, then the maximum voltage at which the cable components may work is the usual limitation on the number of repeaters. The power supply is usually fed from both ends with constant current sources and use the sea as a return path. The exact voltage/current supply characteristic of the power sources is tailored to produce the same magnitude of voltage at either end.

Bandwidth conservation
On long distance systems it is economic to provide more complicated terminal equipment in order to provide a higher utilisation of the system. The first economy possibility is in bandwidth. Since speech occupies only 3·1 kHz (300 to 3400 Hz) then using 4 kHz channels means that 22·5 per cent of the bandwidth is wasted. In order to reduce this, 3 kHz channel equipment has been developed which transmits audio frequencies of 200 Hz to 3050 Hz, i.e. 2·85 kHz bandwidth which only wastes 5 per cent. In order to achieve such close packing it is necessary to filter the audio signal before it is modulated. It is then possible to produce 16 channel groups in the same band as a conventional 12 channel, 4 kHz spacing. The reduction in the upper audio frequency transmitted is considered worthwhile for the 33 per cent increase in speech channels produced.

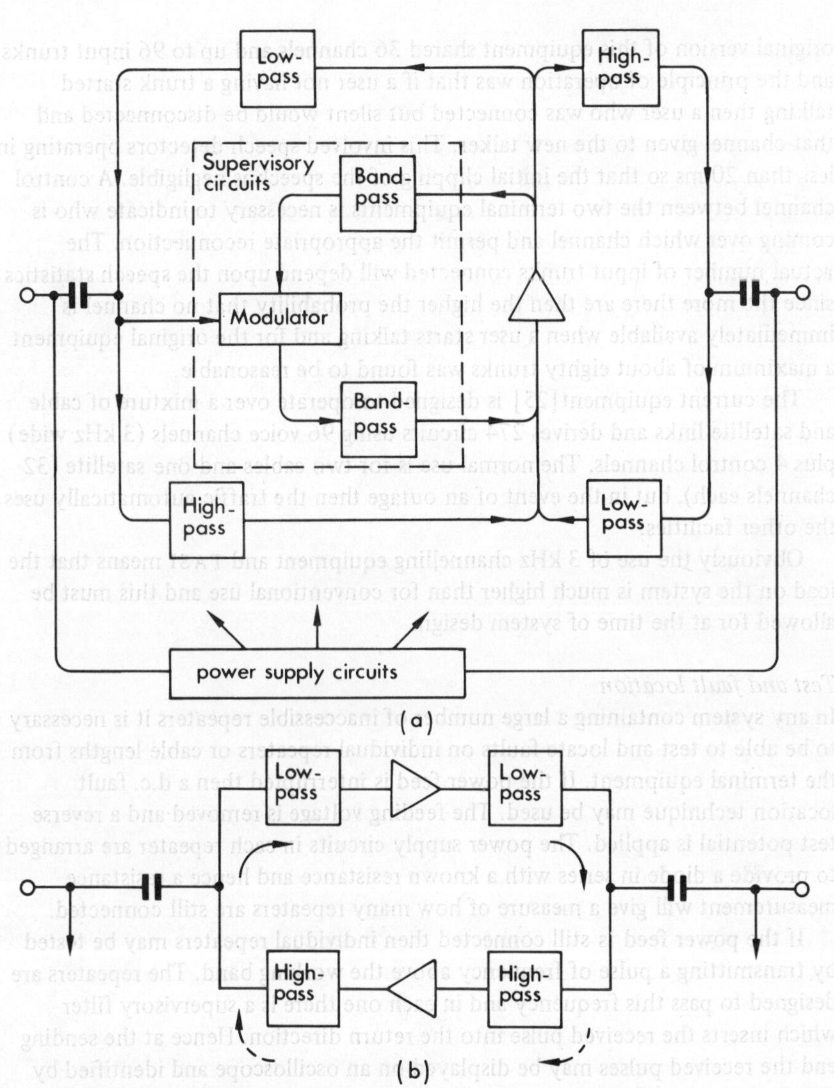

Figure 4.12 Submarine repeater arrangement
(a) Single amplifier, (b) double amplifier for wide-band systems

The other area of economy is in time. In any conversation each participant speaks for only 40 per cent of the time. The rest of the time is spent listening and waiting for replies. Also, even on a highly loaded route during the busy hour, each channel is not used full-time, since to provide a high probability that a circuit is available on demand then the *average* occupancy is only around 80-90 per cent. This spare time may be utilised by providing high-speed audio switching and only give a person the use of a channel when he needs it. The equipment to achieve this is called TASI (Time Assignment Speech Interpolation)[24]. The original version of this equipment shared 36 channels and up to 96 input trunks, and the principle of operation was that if a user not having a trunk started talking then a user who was connected but silent would be disconnected and that channel given to the new talker. This involved speech detectors operating in less than 20 ms so that the initial clipping of the speech is negligible. A control channel between the two terminal equipments is necessary to indicate who is coming over which channel and permit the appropriate reconnection. The actual number of input trunks connected will depend upon the speech statistics since the more there are then the higher the probability that no channel is immediately available when a user starts talking and for the original equipment a maximum of about eighty trunks was found to be reasonable.

The current equipment[25] is designed to operate over a mixture of cable and satellite links and derives 274 circuits using 96 voice channels (3 kHz wide) plus 4 control channels. The normal use is for two cables and one satellite (32 channels each), but in the event of an outage then the traffic automatically uses the other facilities.

Obviously the use of 3 kHz channelling equipment and TASI means that the load on the system is much higher than for conventional use and this must be allowed for at the time of system design.

Test and fault location

In any system containing a large number of inaccessible repeaters it is necessary to be able to test and locate faults on individual repeaters or cable lengths from the terminal equipment. If the power feed is interrupted then a d.c. fault location technique may be used. The feeding voltage is removed and a reverse test potential is applied. The power supply circuits in each repeater are arranged to provide a diode in series with a known resistance and hence a resistance measurement will give a measure of how many repeaters are still connected.

If the power feed is still connected then individual repeaters may be tested by transmitting a pulse of frequency above the working band. The repeaters are designed to pass this frequency and in each one there is a supervisory filter which inserts the received pulse into the return direction. Hence at the sending end the received pulses may be displayed on an oscilloscope and identified by their time of arrival. This permits the testing of a repeater and an indication of its gain.

In a submarine system more information is needed of the behaviour of noise and gain in individual repeaters and this is provided by continuous tone testing[26]. This is normally operated from the cable end which uses the high-band for the GO directions. The principle of operation is to use a specific combination of two test frequencies in the high-band to select a repeater, and produce a return signal in the low-band which is then transmitted back to the terminal. This is achieved by providing a modulator in each repeater where the carrier is obtained via a narrow band-pass filter from the output of the common amplifier and whose input comes from the GO direction. The output of the modulator is passed via another narrow band filter to the input of the common amplifier. Hence each repeater will only respond to one combination of carrier and test frequency. This allows the measurement of the loop gain of all the repeaters from the terminal to the selected repeater. If only one carrier frequency is sent to line then the supervisory unit will continue to return signals which will be the result of noise at the test frequency. Hence a measure of the noise may also be computed.

REFERENCES

1. A. H. Flores and T. L. Moore, 'The Evolution of Station Carrier and Recent Operating Experience', *Trans. I.E.E. Comm. Tech.* **COM-19**, April 1971, pp. 211-217.
2. *C.C.I.T T. White Book III*, Recommendation G211 and G233.
3. H. B. Law *et al.*, 'Channel Equipment Design for Economy of Bandwidth', *Post Office Electrical Engineers Journal* **53**, 1960, p. 112.
4. *C.C.I.T.T. White Book III*, Recommendation G135.
5. A. F. G. Allan, 'Small-Diameter Coaxial Cable Developments', *Post Office Electrical Engineers Journal* **57**, 1, April 1964, p. 1.
6. For a detailed discussion, see *Transmission Systems for Communication*, Chapter 11, Western Electric, 1970.
7. *Ibid*, Chapter 10.
8. W. R. Bennett, 'Cross-Modulation in Multi-Channel Amplifiers Below Overload', *B.S.T.J.* **19**, October 1940, pp. 587-610.
9. R. A. Brockbank and C. A. Wass, 'Non-Linear Distortion in Transmission Systems', *Journal I.E.E.* **92**, III, 1945, p. 45.
10. J. C. H. Davies, and H. O. Freidheim, 'Intermodulation on Amplitude-Modulated Multi-Channel Line Link', *Proc. I.E.E.* **107C**, June 1960, pp. 342-52.
11. B. S. Helliwell *et al.*, 'The Design of a Wideband Transistor Telephone Repeater and its Distortion Performance', *Proc. I.E.E.* **107C**, December 1961, pp. 230-5.
12. J. L. Garrison *et al.*, 'Basic and Regulating Repeaters', *B.S.T.J.* **48**, 4, April 1969, pp. 841-88. This is a complete issue devoted to the L-4 coaxial system.
13. R. W. Berry, 'Speech Volume Measurements on Telephone Circuits', *Proc. I.E.E.* **118**, February 1971, pp. 335-338.
14 For example, see W. R. Bennett, 'Cross-Modulation Requirements on Multi-Channel Amplifiers Below Overload, *B.S.T.J.* **19**, October 1940, pp. 587-610.
15. *C.C.I.T.T. White Book III*, Annex 6.
16. K. L. McAdoo, 'Speech Volumes on Bell System Message Circuits—1960 Survey', *B.S.T.J.* **42**, 1963, pp. 1999-2012.
17. *C.C.I.T.T. White Book III*, Recommendation G223.
18. B. D. Holbrook and J. T. Dixon, 'Load Rating Theory for Multi-Channel Amplifiers', *B.S.T.J.* **17**, 1938, pp. 624-44.
19. D. L. Richards, 'Statistical Properties of Speech Signals', *Proc. I.E.E.* **111**, May 1964, pp. 941-949.
20. *C.C.I.T.T. White Book III*, Recommendation G223.

21. *Transmission Systems for Communication*, Chapter 14, Western Electric, 1970.
22. *C.C.I.T.T. White Book III*, Recommendation G212.
23. For example, see the articles on the Bell L-4 System in *B.S.T.J.* 48, April 1969.
24. C. E. E. Clinch, 'Time Assignment Speech Interpolation (TASI)', *Post Office Electrical Engineers Journal* 53, 1960, pp. 197-200.
25. G. R. Leopold, 'TASI-B A System for Restoration and Expansion of Overseas Circuits', *B.L.R.* 48, November 1970, pp. 299-306.
26. F. Scowen, 'Location of Faults on Submarine Telephone Cables', *Post Office Electrical Engineers Journal* 54, January 1962, p. 252.

Chapter 5

Propagation

5.1 Introduction

The previous chapters have essentially been concerned with closed forms of transmission as distinct from free space transmission utilising radio waves. This chapter deals with the effect of the earth and its surrounding atmosphere on the propagation of such radio waves.

The factors affecting the propagation are so numerous and diverse in nature, and depend so much on meteorological and extra-terrestrial phenomena, that no concise theory of propagation exists. The practising engineer relies very much on the observed effects of previous systems and on the exhaustive data supplied from meteorological sources. Theory is thus supplemented to a large extent by charts and routine procedures based upon accumulated practical experience.

The main consideration herein is the effects of the medium upon propagation in a fairly physical, non-mathematical form, with the express aim of deriving the fundamental limitations which govern the choice of operating frequency of a specific communication link, and the practical constraints on systems operation. The various means of propagation can be separated fairly readily into the existing internationally agreed frequency bands, as shown in Table 5.1. It must be noted that this is a fairly arbitrary frequency separation and some mechanisms do overlap the frequency bands designated.

The international allocation of frequency bands for services and areas is regulated by the Administrative Radio Conference of the ITU in Geneva. This conference is held, as required, at intervals of ten to fifteen years, the last one being convened in 1959, and the currently valid frequency allocations, agreed at that time, are contained in Radio Regulations [1] (additions were added in 1963 to accommodate satellite communications [2]). The latest conference took place in 1971. The vast majority of the world's administrations recognise the Radio Regulations and channel arrangements based on them and recommended by the CCIR [3] as binding rules for international radio communication, and also use these rules for the allocation of specific national services. In this way, interference both nationally and across frontiers, can be minimised, and interconnection of national systems is possible (i.e. the Eurovision link).

Before progressing to discuss the propagation mechanisms more deeply, it

Table 5.1 Regions of Propagation

Frequency	Band	Propagation Mechanism	Systems
3 Hz → 3 kHz	e.l.f.		No uses
3 kHz → 30 kHz	v.l.f.	Earth-Ionosphere waveguide	Worldwide, military and navigation
30 kHz → 300 kHz	l.f.	Surface wave	Long wave band. Stable transmission, distances up to 1500 km
300 kHz → 3 MHz	m.f.	Surface wave (short distances) Sky wave (long distances)	Medium wave band broadcasts. Sky wave over longer distances–subject to fading.
3 MHz → 30 MHz	h.f.	Sky wave	Short wave communication band 3 – 6 MHz – continental 6 – 30 MHz – intercontinental inside skip distance short distance ground waves possible mobile land communication ship-shore communication
30 MHz → 300 MHz	v.h.f.	Space wave	line of sight v.h.f. links over short distances
		Scatter wave	> 50 MHz ionospheric scatter 2000 km., over long distances
300 MHz → 3 GHz	u.h.f.	Space wave	line of sight, short distances
		Scatter wave	> 500 MHz upwards tropospheric scatter over long distances–600 km
3 GHz → 30 GHz	s.h.f. (microwave)	Space wave	Space vehicle communication satellite links microwave radio relays etc.
30 GHz → 300 GHz	e.h.f. (millimetre wave)		Line of sight millimetre links, satellite links and possible local distribution systems

will be instructive to consider, from a traffic point of view, the historical usage of the bands.

In the early days of radio, worldwide communications were developed in the long wave or v.l.f. band, because attenuation of radio waves in the atmosphere is proportional to frequency, and this necessitated the use of the lowest practical frequencies. Unfortunately, such systems had the disadvantage of large radiators and high transmitting powers, and also that at the lower frequencies atmospheric noise was a limit to reception sensitivity.

At the turn of the century, Lodge demonstrated the propagation of short waves, and this was further supported by Marconi who succeeded in transmitting transoceanic radio signals. This, together with the independent discovery of a reflecting region in the earth's upper atmosphere by Kenelly and Heaviside, known as the ionosphere, saw the birth of h.f. communications. Since Marconi's development of the basic techniques of h.f. propagation, for forty years the bulk of long distance traffic throughout the world was carried by h.f. signals propagated in the ionosphere. There were many reasons for such heavy reliance on h.f. radio; it was inexpensive in comparison with other methods and was the most reliable and fastest means of communication possible.

The rapid advances in v.h.f. and u.h.f. devices resulting from second world war radar achievements gave a tremendous impetus to the v.h.f. line-of-sight systems being studied just prior to the war. Commercial pressure for more bandwidth resulted in such radio relay systems springing up in both the United Kingdom and the United States of America. The birth of television in the 1940's and continuing pressure for more telephone circuits saw these systems steadily move into the u.h.f. part of the spectrum, and their numbers increased drastically. Such microwave radio relays, together with the more recent introduction of satellite relays in the 1960's (which are merely a large scale version of the terrestrial relay) and the troposcatter relay, made possible by the discovery of tropospheric scattering in the 1950's, form today's major traffic-carrying routes. Far from making lower frequency communications obsolete, the advances have freed them to do the job for which they are best suited, i.e. low frequencies for worldwide and military communications and navigation, medium frequencies predominantly for broadcasting and high frequencies for carrying low and medium-density message traffic over moderately long paths and at low costs. In addition, these bands, particularly the high frequency, have assumed a new role as back-up for the higher density, higher frequency routes, where any failure in the primary system interrupts service.

5.2 Methods of propagation

Simplest propagation is that in free space, which cannot be achieved in practice of course, due to the presence of the earth and the imperfectness of the atmosphere, but which gives useful reference for actual conditions. The free space attenuation between two isotropic radiators spaced a distance d apart and radiating energy of wavelength λ is

$$20 \log_{10}\left(\frac{\lambda}{4\pi d}\right) \text{dB} \tag{5.1}$$

In general, this attenuation will be modified by the properties of the atmosphere through which, or the ground over which, the radio waves are transmitted, and this usually takes the form of correction factors which are added to (5.1). Radio waves are transmitted in free space so that their wavefronts are spherical and have radii of curvature which are continually expanding. Such a wave can be assumed to travel in a straight line as indicated in Figure 5.1. The principle of ray-optics can thus be used to determine its propagation course, but one must not lose sight of the actual spherical nature of a wavelet. (N.B. free space attenuation is due to the expansion of the wavefront and the resultant distribution of power over a larger area).

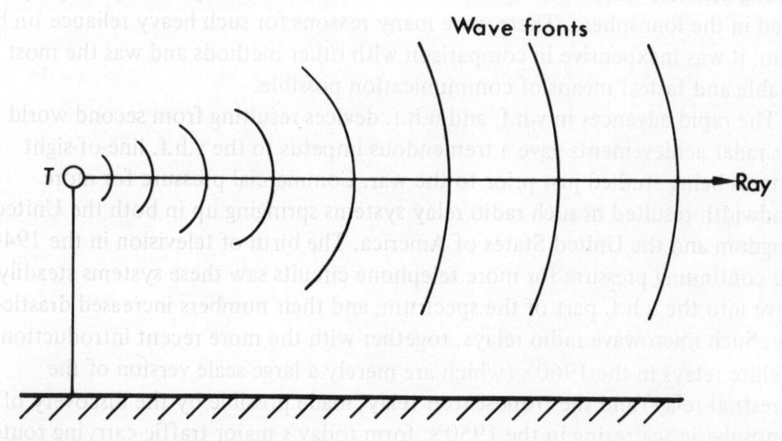

Figure 5.1 Ray optical theory

If we now consider the practical cases of terrestrial propagation, it will be obvious from Figure 5.2 for a transmitter and receiver situated on the earth's surface that if a direct line can be drawn between them, a signal will be received. In addition to this direct wave there can exist a reflected wave arriving at the receiver after reflection from the earth's surface. These make up the *space wave*, i.e.

space wave = (direct + ground reflected) wave

If, however, the separation between transmitter and receiver is such that the curvature of the earth precludes line of sight, then no space wave is received. However, propagation is still possible beyond the horizon due to two important means which are respectively *surface wave* and *sky wave*. The surface wave is produced by energy travelling close to the ground and guided by it to follow the

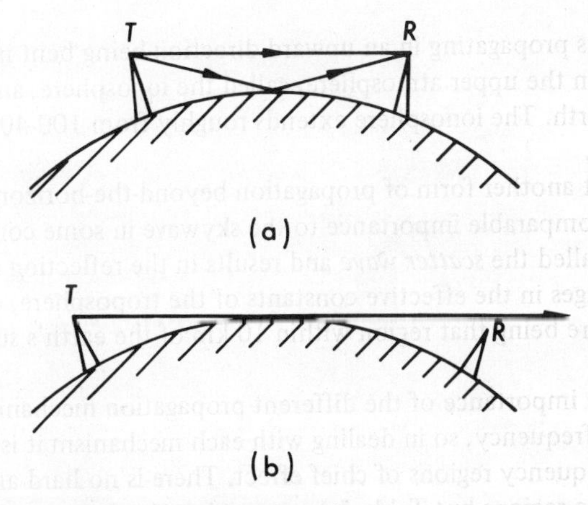

Figure 5.2 Ray paths over a curved earth
(a) within optical horizon (b) beyond optical horizon

curvature of the earth much as an electromagnetic wave is guided by a transmission line. This is the phenomenon of diffraction.

The direct, reflected and surface (or diffracted) wave may all be present together, and it is not always necessary to separate them. The combination is known as the *ground wave*.

ground wave = surface wave + space wave

The sky (or ionospheric) wave, as can be seen from Figure 5.3, is dependent

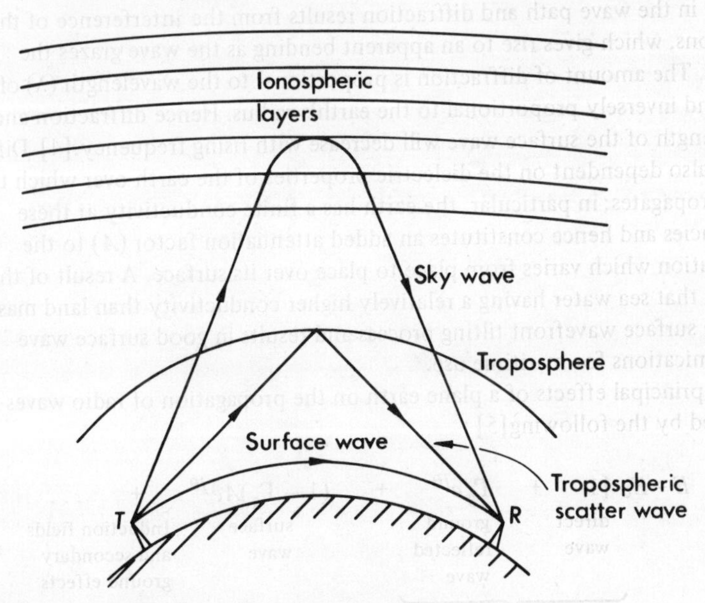

Figure 5.3 Beyond the horizon propagation

upon the waves propagating in an upward direction being bent in an electrically charged layer in the upper atmosphere, called the ionosphere, and reflected back towards the earth. The ionosphere extends roughly from 100-400 km above the earth's surface.

There is yet another form of propagation beyond-the-horizon which is becoming of comparable importance to the skywave in some communication links. This is called the *scatter wave* and results in the reflection of waves due to turbulent changes in the effective constants of the troposphere, or ionosphere. The troposphere being that region within 10 km of the earth's surface as shown in Figure 5.3.

The relative importance of the different propagation mechanisms is strongly dependent on frequency, so in dealing with each mechanism it is useful to bear in mind the frequency regions of chief effect. There is no hard and fast boundary between the regions but Table 5.1 summarises the frequency dependence of the mechanisms and some of their main uses.

5.3 Low-frequency propagation

The means of propagation in this part of the spectrum is via surface waves which result from the diffractive bending of the wave motion around obstacles (cf. sound waves); in this case the obstacle is the earth's surface.

The phenomenon of diffraction may be understood if one considers the primary wavefront to be made up of an infinite number of secondary isotropic sources. When an obstacle is met, the secondary sources radiate from its entire surface in the wave path and diffraction results from the interference of their radiations, which gives rise to an apparent bending as the wave grazes the surface. The amount of diffraction is proportional to the wavelength (λ) of the wave and inversely proportional to the earth's radius. Hence diffraction and thus the strength of the surface wave will decrease with rising frequency.[4] Diffraction is also dependent on the dielectric properties of the earth over which the wave propagates; in particular, the earth has a finite conductivity at these frequencies and hence constitutes an added attenuation factor (A) to the propagation which varies from place to place over its surface. A result of the latter is that sea water having a relatively higher conductivity than land masses, aids the surface wavefront tilting process and results in good surface wave communications for maritime use.

The principal effects of a plane earth on the propagation of radio waves is indicated by the following[5] :

$$E = E_0\,[\underbrace{\underset{\text{direct wave}}{1} + \underset{\text{ground reflected wave}}{\Gamma_g e^{j\theta}}}_{\text{space wave}} + \underset{\text{surface wave}}{(1-\Gamma_g)Ae^{j\theta}} + \underset{\text{Induction fields and secondary ground effects}}{\ldots} \tag{5.2}$$

where E_0 is the direct (line of sight) field strength
Γ_g is the reflection coefficient of the ground which depends on polarisation [6]
θ is the path phase difference between direct and reflected waves,
A is the surface wave attenuation factor depending on antenna separation frequency and the earth's characteristics [6].

It is found that for large separation distances and low antenna heights, i.e. small angles of incidence, the reflection coefficient of the earth is independent of polarisation and approximately equal to 1. As the path lengths are almost equal, this means that the two components of the space wave practically cancel and the surface wave is the predominant one in the received field. In fact, surface wave propagation is only successful when the antennas are small compared to a wavelength and located near to the ground at frequencies less than 1 MHz. If the antennas are elevated above the earth's surface, the space wave components do not cancel and this becomes the predominant received wave as the surface wave is rapidly attenuated. (For this reason v.h.f./u.h.f. antennas are usually elevated).

Low-frequency radio spectrum usage

The low frequency spectrum is mainly used for worldwide navigation systems; for instance the lowest frequency in regular use is 10·2 kHz, by the US Navy Omega long distance navigation system. This v.l.f. band is reserved for navigational systems [1] and in general uses vertically polarised monopoles transmitting powers in the megawatt range.

In the v.l.f., band frequencies are encountered such that the earth's surface to ionosphere height becomes comparable to a wavelength, when it is no longer valid to employ simple ray-optical concepts as described earlier. Under these conditions it has been found more desirable to consider the propagation mechanism as one in which the radio waves are guided between the ionosphere and the earth's surface in terms of waveguide theory [7]. It has been shown that this approach leads to valid results for worldwide propagation in the v.l.f. band [8] where the earth and ionosphere appear as conducting planes to the waves, which then propagate in the so called earth-ionosphere waveguide mode. Using this method of propagation at 10 kHz, worldwide communication to submerged submarines should be possible, although the problems involved are still as yet largely unsolved [9].

In the l.f. band exist telegraph services, both morse and teleprinter, as well as maritime and aeronautical beacon, weather forecast and status information bands which do in fact just extend into the m.f. main broadcast band [10].

5.4 High-frequency propagation

As was seen earlier, sky waves result from reflections off the ionosphere and thus provide long-distance over-the-horizon communication. In general, the 'sky wave' signals are less stable than the ground wave signals, their strength depending upon the frequency, and upon the conditions of the ionosphere. The state of the ionosphere is found to vary from hour to hour, day to day, and season to season in much the same way as the weather. Ionospheric forecasting is a major part of sky wave propagation and worldwide stations have been set up to record information on the ionosphere which is made available in the form of charts that show past conditions and also make predictions for the future to aid prospective users. Using these charts, it is possible to determine in advance the optimum frequency to use for communications between any two points on the earth's surface at any given time.

The ionosphere

The ionosphere is formed by the sun's ultra-violet and X-radiation ionising the molecules of the upper atmosphere. Although ions and electrons are undoubtedly present throughout the whole region which lies between 250-350 km above the earth's surface, it has been found that there are several layers in which the ionisation density reaches a maximum. These layers are designated by letters *D*, *E* and *F* in order of height. At times the *F* layer splits into separate layers called *F*1 and *F*2. The actual distribution of ionisation density varies according to time of day, month, year and geographical location. A typical electron density height profile is shown in Figure 5.4. There are in general two effects on

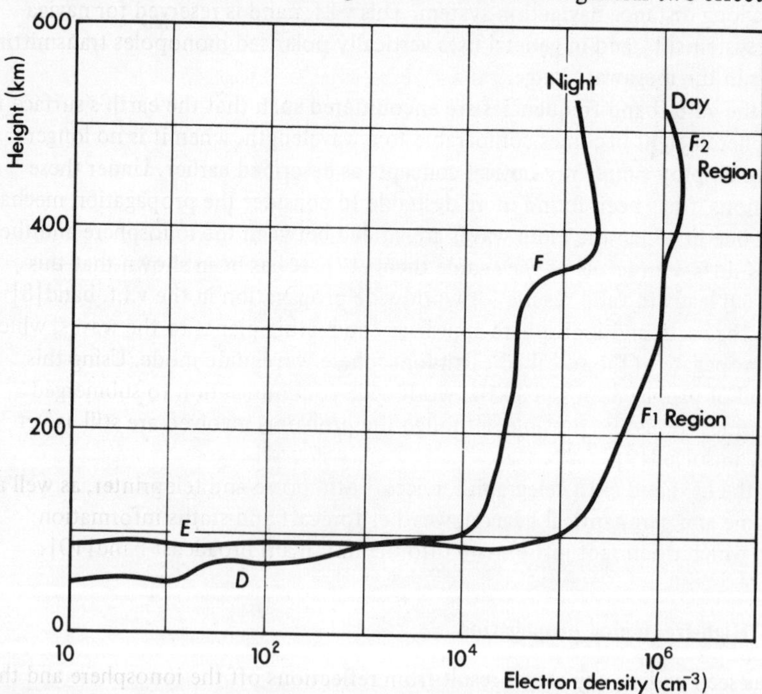

Figure 5.4 Ionospheric electron/ion density–height profile (from ref. 11)

radio waves in the ionosphere:

(a) Refractive bending. Due to the refractive index changing with the degree of ionisation, frequencies below a critical wave value may be bent so much as to return to earth. (See Figure 5.5). This phenomenom is associated with the E and F layers.

(b) Absorption. Free electrons are set in motion by the electric field of radio waves which gives rise to an interchange of energy from the

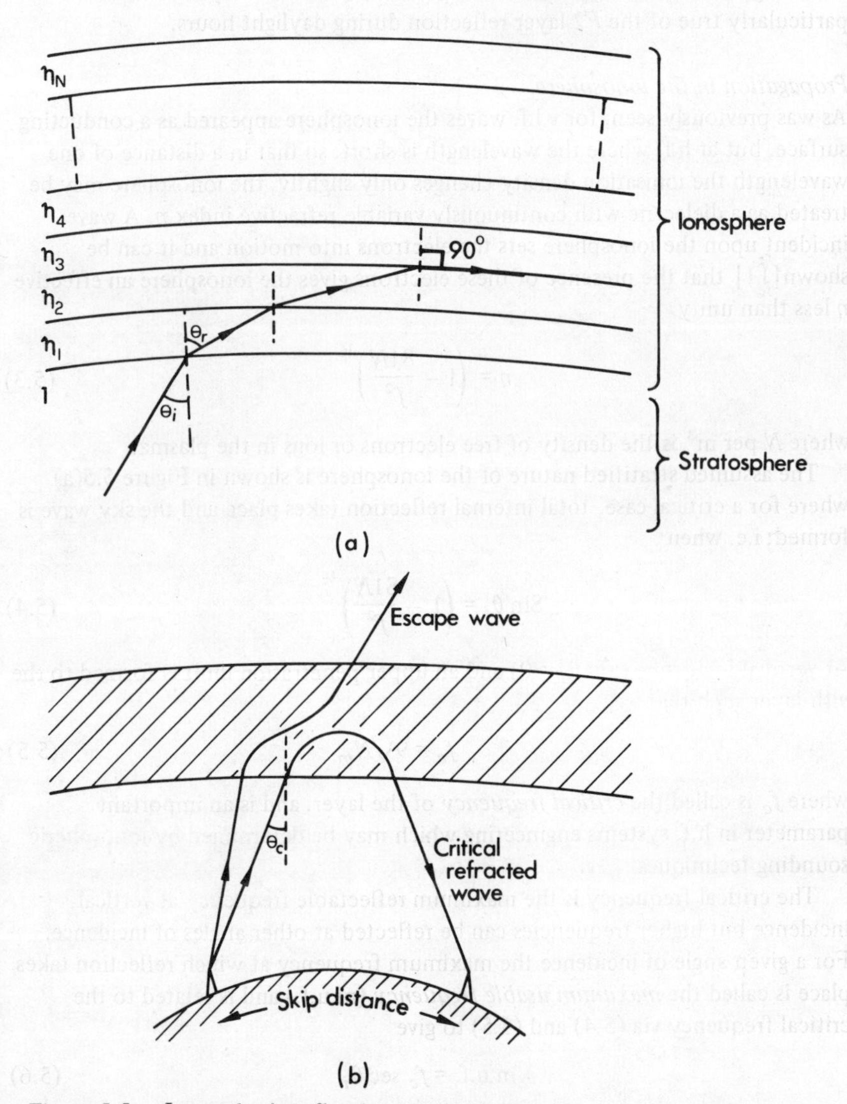

Figure 5.5 Ionospheric reflection mechanism
(a) layer refraction (b) critical conditions for reflection

waves when an electron collides with a molecule. Absorption is thus greatest in the region of higher molecular density, i.e. D, and much lower in E and F layers. Waves are attenuated in proportion to $1/f^2$; hence the greater the frequency the less the attenuation. Attenuation is also greatest during daytime due to the existence of the D layer which disappears at night due to molecular recombination.

The refractive or bending properties increase with ionisation density and thus with height, i.e. longer distance communications are via F layers, and this is particularly true of the $F2$ layer reflection during daylight hours.

Propagation in the ionosphere

As was previously seen, for v.l.f. waves the ionosphere appeared as a conducting surface, but at h.f. where the wavelength is short, so that in a distance of one wavelength the ionisation density changes only slightly, the ionosphere may be treated as a dielectric with continuously variable refractive index η. A wave incident upon the ionosphere sets the electrons into motion and it can be shown[11] that the presence of these electrons gives the ionosphere an effective η less than unity

$$\eta = \left(1 - \frac{81N}{f^2}\right)^{1/2} \tag{5.3}$$

where N per m^3 is the density of free electrons or ions in the plasma.

The assumed stratified nature of the ionosphere is shown in Figure 5.5(a) where for a critical case, total internal reflection takes place and the sky wave is formed; i.e. when

$$\text{Sin}\ \theta_i = \left(1 - \frac{81N}{f^2}\right)^{1/2} \tag{5.4}$$

At vertical incidence (Sin $\theta_i = 0$) and an upper penetration limit is formed to the mth layer such that

$$f_c = 9\sqrt{N_m} \tag{5.5}$$

where f_c is called the *critical frequency* of the layer, and is an important parameter in h.f. systems engineering which may be determined by ionospheric sounding techniques.

The critical frequency is the maximum reflectable frequency at vertical incidence but higher frequencies can be reflected at other angles of incidence. For a given angle of incidence the maximum frequency at which reflection takes place is called the *maximum usable frequency* (m.u.f.) and is related to the critical frequency via (5.4) and (5.5) to give

$$\text{m.u.f.} = f_c \sec \theta_i \tag{5.6}$$

This is the maximum communications frequency for reflection from a particular

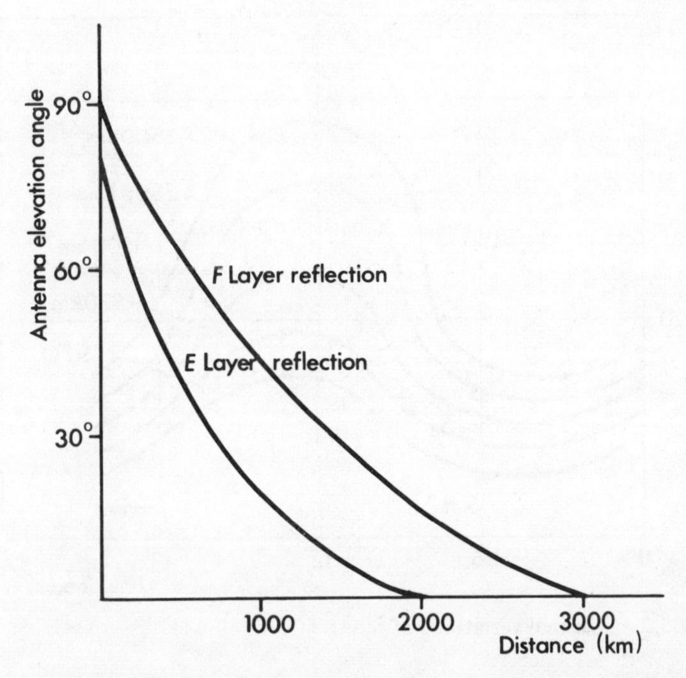

Figure 5.6 Ionospheric transmission distance as a function of antenna beam elevation

layer of critical frequency f_c. At the m.u.f. the wave path is that of the critical path (see Figure 5.5b) and the transmission distance is thus the minimum distance at which the sky wave reaches the earth again; this is called the *skip distance.*

The maximum transmission distance against elevation angle of the antenna beam is shown in Figure 5.6, for the two layer E, F standard heights. N.B. In practice angles too near to zero are avoided as the wave suffers considerable ground loss. The lowest usable angle is usually 5°. The largest angle of incidence obtainable with F layer reflection is of the order of 74°, where here the m.u.f. = $3{\cdot}6\, f_c$.

As attenuation is proportional to $1/f^2$, the strongest sky wave signals at the receiver are found at the highest possible working frequency, and hence it is necessary to work as near to the m.u.f. as possible.

Figure 5.7 shows a set of m.u.f. predictions for a typical day. Such charts are issued by several organisations[12] giving monthly median predictions of m.u.f. Variations of m.u.f. occur diurnally and these change from day-to-day over a month. The median value is only statistically reliable for 50 per cent of the days; this is clearly not acceptable in practice as it would produce deviations of up to 15 per cent on the median value. Thus it is usual to use a frequency somewhat lower than the predicted m.u.f. (actually 15 per cent below it) which we call the

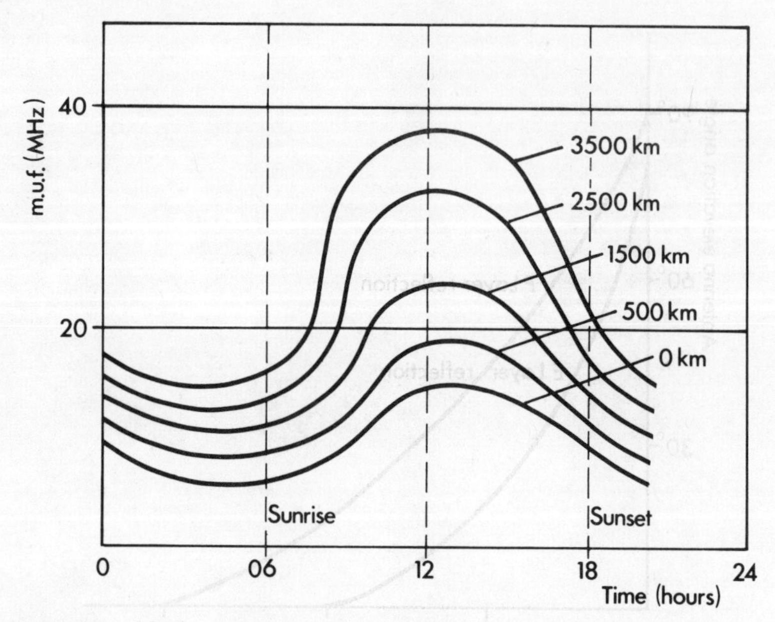

Figure 5.7 Diurnal variations of m.u.f. (from ref. 11)

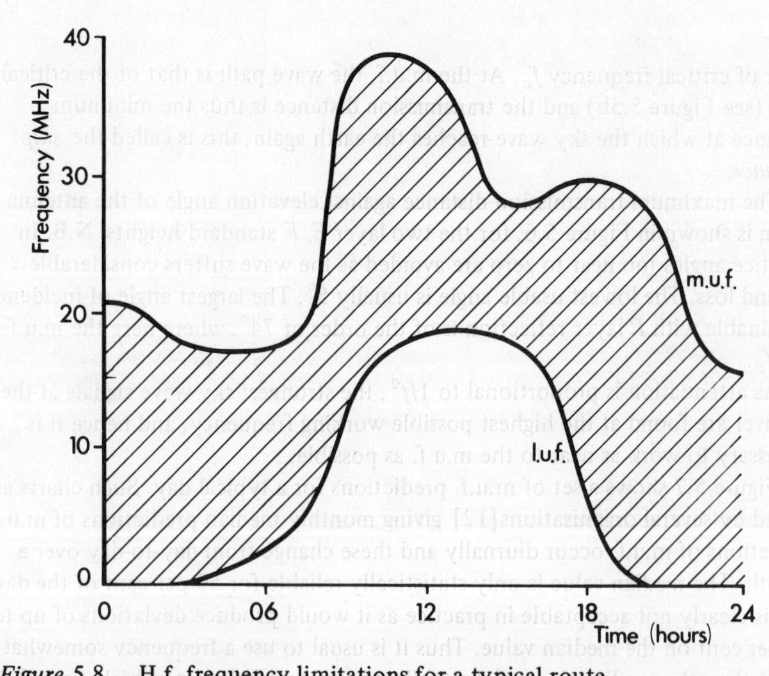

Figure 5.8 H.f. frequency limitations for a typical route

optimum working frequency (o.w.f.),

$$\text{o.w.f.} = 0{\cdot}85 \text{ m.u.f.} \quad (5.7)$$

such that it is less than the m.u.f. for 90 per cent of the time and variations have a normal distribution (standard deviation of 0·117 times m.u.f.).

Working below the m.u.f. has the disadvantage of introducing more attenuation due to extra absorption losses[13]. For a given communication link, if transmitter power, total antenna gain and noise background at the receiver are known, it is possible to deduce the lowest usable frequency (l.u.f.) in order to exceed a minimum acceptable signal-to-noise ratio at the receiver[14]. However, all such methods employ empirically determined constants and are at present under review by the CCIR [15].

Frequency allotment

The limitations placed on the choice of frequencies for a typical long distance link are shown in Figure 5.8. The spectrum of usable frequencies is extremely small, thus making their allocation a very complex problem. Frequency sharing on an international basis provides a limited solution to this in the crowded h.f. band[1]. Individual transmitters are usually allocated several spot frequencies to operate at daytime and night-time (between which rapid switching is necessary) due to the diurnal variation shown in Figure 5.8 to maintain a reliable service.

Multiple-hop transmission

For distances greater than 4000 km, which is about the maximum distance of transmission of $F2$ layer reflection, transmission can take place by multiple hops, i.e. by successive reflections at the ionosphere and at the earth's surface. This form of transmission is not a simple extension of the single hop case due to the diffusion of the waves on successive journeys through the ionosphere. The waves arrive at the receiver via a number of different routes and it becomes difficult to ascribe any one preferred angle of elevation for the receive antenna. Diversity systems are often used in such cases, where two or more antennas are spaced a given distance apart in the case of space diversity, the theory being that they will not all experience fades at the same time and thus switching may be used to select the one with the best signal-to-noise received. Frequency diversity is also used, especially to combat frequency selective fading, but more successfully in practice, a combination of the two called quadruple diversity is used in which the advantages of frequency diversity may be improved by using two or more antennas, each connected to a separate receiver and an amplitude comparator to select the larger output.

Ionospheric predictions

Determination of the optimum working frequency for a particular link has in the past been reduced to operations on 'ionospheric forecasts' which are issued several months in advance. These forecasts being deduced from previously measured values and issued in the form of world contour charts of either f_c or m.u.f., whence graphical techniques could be used to determine the m.u.f. between two particular points on the chart (for a description of the techniques see pp. 458-461, reference 11). Alternatively, for the amateurs, Wireless World publishes monthly forecasts of m.u.f. and l.u.f. for communication between major cities.

A more recent approach recommended by the CCIR [16] obviates the necessity for contour charts and relies only on the forecasts of a quantity called the *ionospheric index* (denoted I_{F2} for the $F2$ layer) (for a complete description, see Chapter 10, reference 17). This method was developed by noticing that,

(a) there was a linear relationship between f_c as observed between any two points on the earth's surface, and
(b) that there was a linear relationship between corresponding hourly values for the same month in consecutive years.

It was concluded that forecasts of f_c giving worldwide coverage, could be evaluated from predictions made for one location only by interpolation from previously recorded data already available in world contour chart form. Instead of issuing charts, the issue of a single factor having a linear relationship to the monthly median value of f_c at a reference location would suffice. This factor is the ionospheric index I, forecasts of which appear in Telecommunications Journal monthly, and together with standard data [18], enable f_c to be calculated at the reflection points of any communication link and thus m.u.f. and o.w.f. found.

The present method of prediction is more accurate in that it looks at the current trend in I and compares it with previous years' trends, both hourly for consecutive years and at corresponding sunspot cycle points, whereas the previous method relied on predictions six months in advance. However, pessimistic forecasts are still the rule, and many links suffer from over engineering and waste valuable spectrum. Continuous pulse sweeping of the transmitter and a synchronously locked receiver may be the ideal answer, as this will enable an on-line ionogram and accurate frequency selection to be made for each link.

H.f. planning

A review of h.f. equipment and station planning follows in the next chapter; the following is a summary of the propagation planning/procedure:

(i) Determine great circle distance between stations.
(ii) Depending on distance and ionospheric layer to be used, determine number of hops and reflection points.
(iii) Determine critical frequency (f_c) at hop points from forecasts:– m.u.f. = $f_c \sec \theta_i$, o.w.f. = 0·85 m.u.f. and calculate l.u.f.
(iv) Determine take-off angle of antenna beam and gain required. Select antenna. Estimate receive angle of antenna.

Disturbances in ionospheric propagation

(a) *Regular disturbances* Because of the continual changes in the earth's position relative to the sun, the amount of ultra-violet radiation received at any point in the ionosphere is continuously varying. This gives rise to diurnal and seasonal variations of the ionisation density and hence f_c. A typical diurnal variation was shown in Figure 5.7, where m.u.f. rises to a maximum at mid-day and falls off during darkness hours. For obvious reasons winter values will be lower than summer. There is a longer period variation in the ionosphere which is found to correspond with variations in solar activity. The solar variation is the size and number of sun-spots visible to the earth. The sun-spot cycle has a period of approximately eleven years, and effects the ultra-violet radiation incident upon the ionosphere and hence its f_c values.

A further disturbance is caused by the earth's magnetic field which exerts a force on the electrons set in motion by the h.f. wave in the ionosphere and produces a twisting effect on their paths which splits up the wave into two components. This effect is most severe around the geomagnetic dip equator where, due to the orthogonal polarisation of the components, circular polarisation is normally used.

(b) *Irregular ionospheric disturbances* Sudden ionospheric disturbances (s.i.d.'s) or Dellinger fade-outs, as they are sometimes called, cause complete disappearance of the sky wave lasting from between a few minutes to several hours. The cause is sudden eruptions on the sun, called solar flares which emit large amounts of radiation and hence large increases in the ionisation of the *D* layer. Effects predominant in sun spot maxima are confined to the sunlit earth side and are more intense at low latitudes.

Unpredictable ionospheric storms cause abnormally low receiver powers lasting for several days. Storms occur due to solar flares, but are the result of turbulence caused by high energy particle streams in the ionosphere. For this reason they usually occur some thirty hours after an s.i.d., but severe effects are normally felt only once or twice per sun spot cycle. Higher frequencies are affected greatest and thus communication may be maintained by reducing the operating frequency. The effect is felt equally in the dark and sun-lit areas and is particularly severe near the geomagnetic poles.

(c) *Fading* Interference fading results from fluctuations in the ionosphere causing the received signal to be made up of a number of components whose path difference varies. This can be broken down into a slow fade due to layer variations, which varies diurnally and seasonally, and can be combated by adding a fixed value to the median signal value, and a rapid fade due to random fluctuations of the higher layers causing short term variations of up to 30 dB in signal strength. The latter gives a Rayleigh type distribution depending on circuit length, frequency etc., and circuits must thus be planned on a reliability basis. This can be alleviated to some extent by space diversity.

Polarisation fading due to the superposition of the two split wave components is random in nature and can be alleviated by polarisation diversity, or by use of circular polarisation.

Selective or multipath fading results from different frequency components travelling different path lengths and adding at the receiver with random phase and amplitude. This gives severe distortion of modulated signals and restricts bandwidth. Time differences of between 0·5 and 4·5 ms on typical circuits limit the signalling speeds on telegraph channels to about 100 bauds. This form of fading may be remedied by frequency diversity and working near to the m.u.f. to reduce time delays.

Scattering takes place in the ionosphere in the v.h.f. band between 30-50 MHz, but bandwidth is narrow and only suitable for telegraphy. Meteor trails in the *E* layer have also been used for narrow band reflections.

5.5 v.h.f. and u.h.f. propagation

The sky wave mechanism has an upper limit of approximately 30 MHz and, except under special conditions, signals at v.h.f. and above depend upon the space wave mechanism.

In general, a ground wave set will be received by a terrestrial receiver and, due to the fact that the surface wave suffers considerable attenuation at these frequencies, it is necessary to elevate the antennas by at least several wavelengths above the ground to ensure only space wave propagation. Thus the space wave is used for nearly all line-of-sight type communication links, and the region of propagation is the troposphere. The main influence on the quality of the received signal is the variation of the refractive index with height of the air which is dependent on the meteorology.

Standard atmosphere

A 'standard atmosphere' is an approximation used in radio communications, which assumes stable conditions and constant gradients of pressure, temperature and humidity from the ground to the tropopause, i.e. a constant gradient of radio refractive index η as shown in Figure 5.9. The CCIR have adopted a basic relationship of η with height h (km. above sea level) which is

$$\eta = 1 + 289 \times 10^{-6} e^{-0\cdot136h} \tag{5.8}$$

This continuously variable refractive index effectively means that the velocity of propagation of the radio waves varies with height and thus they will be refracted as they travel through the troposphere. This gives rise to an actual path which is somewhat different from the line of sight path, as shown in Figure 5.10, and the transmission path is effectively increased beyond the optical horizon. An analysis of refraction in the troposphere (such as given in reference [11],

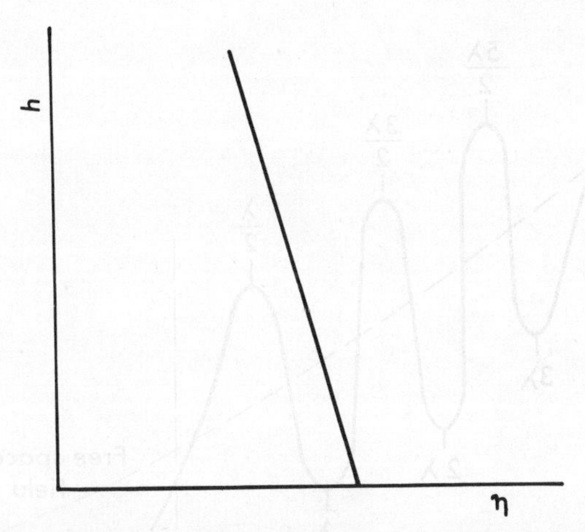

Figure 5.9 Standard atmosphere refractive index profile

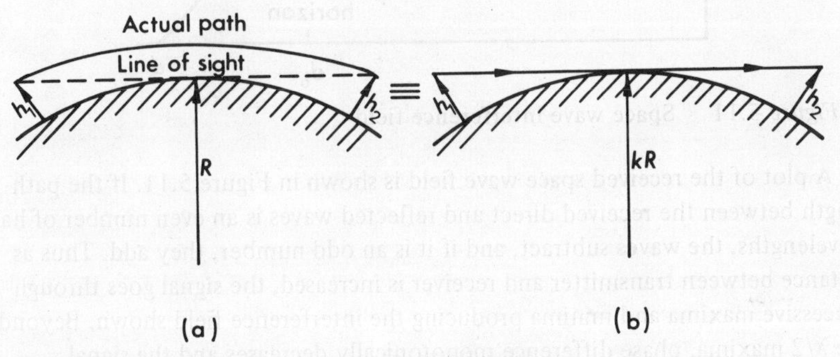

Figure 5.10 Refraction in a standard atmosphere
(a) actual conditions (b) the equivalent, showing modified earth's radius $kR(k = 4/3)$

Appendix 14.6) shows that this effect can be allowed for by using an *effective earth's radius* of kR, where $k = 4/3$ for a standard atmosphere, and treating the waves as propagating over a plane earth.

The space wave comprises both the direct wave, which suffers refraction as discussed, and the ground reflected wave whose magnitude and phase depends on the reflection coefficient of the ground Γ_g , which in turn depends on the ground constants. It also depends on whether vertical or horizontally polarised waves are used. For large distances and small heights, i.e. small angles of incidence, as is usually the case in practice, $\Gamma_g \rightarrow -1$ for both horizontal and vertical polarisations. Reception is also dependent on the roughness of the reflecting ground, i.e. the divergence factor of the ground reflected wave.

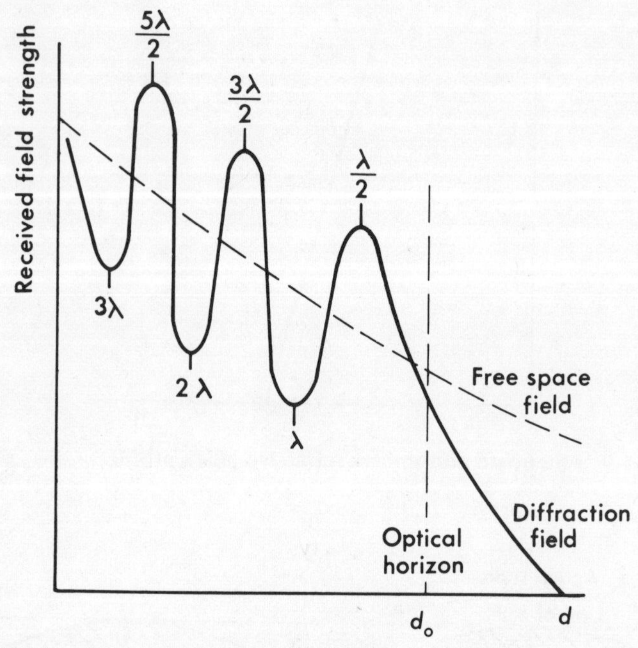

Figure 5.11 Space wave interference field

A plot of the received space wave field is shown in Figure 5.11. If the path length between the received direct and reflected waves is an even number of half wavelengths, the waves subtract, and if it is an odd number, they add. Thus as distance between transmitter and receiver is increased, the signal goes through successive maxima and minima producing the interference field shown. Beyond the λ/2 maxima, phase difference monotonically decreases and the signal approaches zero as the waves destructively interfere.

Non-standard atmosphere

In an atmosphere where the refractive index gradient is changing with altitude, a stratified form may be assumed and propagation over a curved or spherical earth may be considered as over a plane earth by using the *modified refractive index*, rather than a modified earth's radius. The important feature is the curvature of the ray with respect to the earth, so in this case it is modifying η to η_m to reduce to a plane earth, whereas before R was modified to obtain the same result with a purely horizontal beam path. The modified refractive index is defined by the CCIR as

$$\eta_m = \left[\eta - 1 + \frac{h}{R}\right] \times 10^6 \tag{5.9}$$

where in general η changes with temperature gradient, specific humidity or pressure.

A profile of η_m is a graphical cross section of the atmosphere. Any path curvatures determined under standard n_m profile will agree with the effective earth radius of $\frac{4}{3}R$ as shown in Figure 5.12. Also shown are two other common conditions:

(a) Substandard–propagated beam bends upwards due to adverse weather conditions and can cause severe fades

(b) Superstandard–propagated beam bends downwards due again to weather conditions and is intercepted by the earth's curvature; again fading is caused. (United Kingdom links are designed with effective earth radius of 0·7R to combat such sub-standard beam bending).

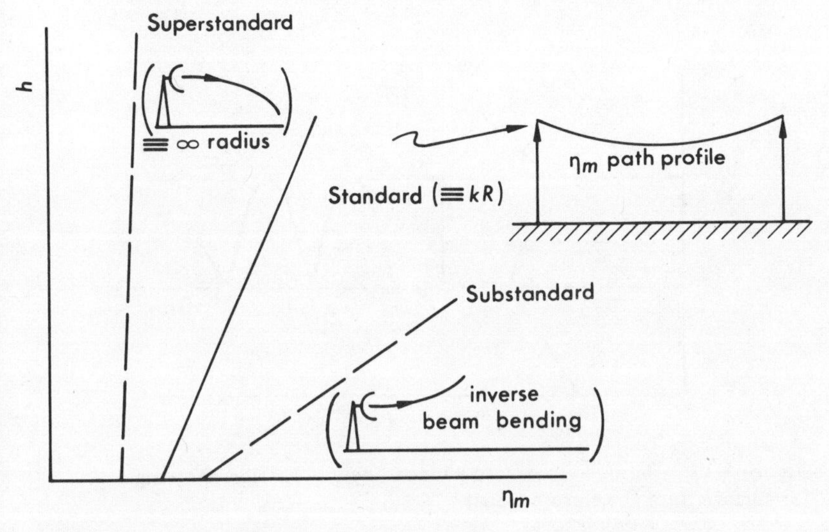

Figure 5.12 Some modified refractive index profiles, showing the equivalent path profile for standard atmosphere case (ray path rather than earth-radius modified)

The atmosphere does not always possess a uniform gradient of η_m, especially under conditions of strong temperature inversion or rapid decrease in water vapour. This gives rise to η_m profiles as shown in Figure 5.13, which refer to conditions of *super-refraction* or *ducting*. This produces a duct either at the surface or elevated levels in which the wave is trapped (it can be thought of as reflected due to the change from sub- to super-standard conditions) and travels as though in a waveguide for long distances around the earth's surface. The wavelength of the trapped wave is dependent upon the height of the duct, the most usual trapping occurring in the microwave region (ducts 10-20 m height). Such non-standard conditions occurring on a link transmission path can cause severe fading and mutual interference with other systems. In certain parts of the world, where a duct is in existence for the majority of the time (i.e. deserts)

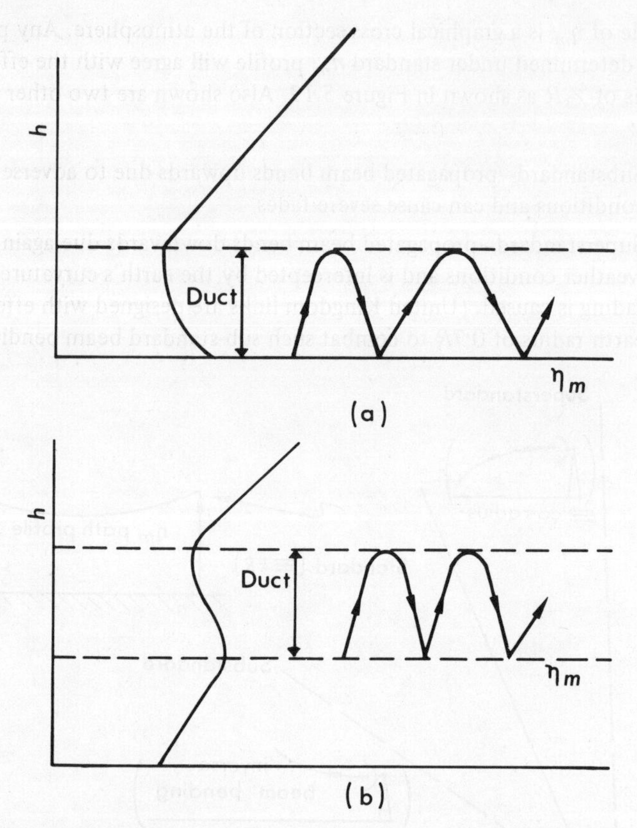

Figure 5.13 Modified refractive index–height profiles showing (a) surface duct (b) elevated duct

propagation can be planned using the phenomena. Due to freak weather conditions, the operational range of existing h.f. communications can be unexpectedly increased by such ducting.

5.6 Microwave link planning

The majority of space-wave line-of-sight systems are to be found in the microwave link networks, which at the moment carry a very large percentage of telecommunications traffic. A typical point-to-point link will have a transmission equation as follows:

$$P_R = P_T + G_T + G_R - L_{TR} - 20\log\left(\frac{4\pi d}{\lambda}\right) \tag{5.10}$$

where P indicates power levels, G the antenna gains of the receiver and transmitter (referred to an isotropic source), and L_{TR} the total equipment losses. The final term in the equation is the free-space loss as given in equation (5.1). This assumes no reflection or refraction of the beam, but a direct path only, and will only apply with accuracy to a properly engineered path.

Path engineering
A properly engineered path is one in which due allowance is made for the clearance of the beam above objects and also its refractive bending.

(a) *Path clearance* At microwave frequencies a careful study of the path profile has to be made before installing a link. Tall buildings, trees, etc., which do not show up on ordnance survey maps, may cause reflections, diffraction or absorption. To avoid this, adequate clearance is necessary for the beam. This is obtained by considering Fresnel interference zones which are cylindrical areas with the direct beam path as centre line, as shown in Figure 5.14. The boundaries between zones are defined by intervals sufficient to increase the direct/reflected path difference by $n\lambda/2$ due to reflections from obstacles. The radius

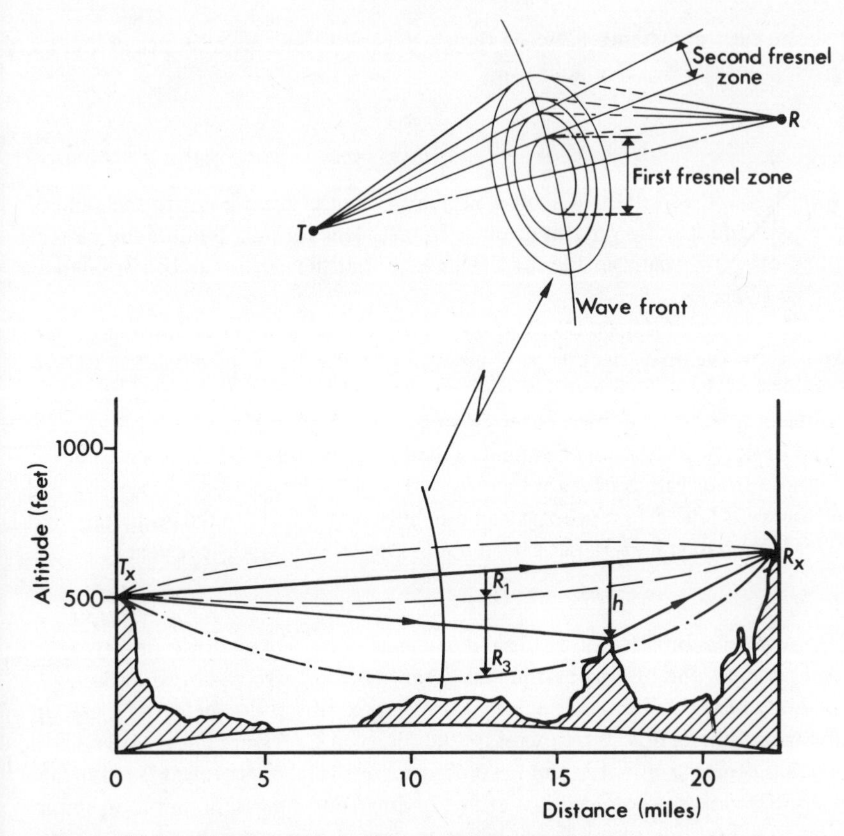

Figure 5.14 Typical path profile showing fresnel zones

of the nth zone at any point along the transmission path may easily be shown to be

$$R_\eta = 13{\cdot}15\sqrt{[\eta\lambda d_1(d - d_1)/d]} \tag{5.11}$$

where d = total path length (miles)

d_1 = distance from transmitter to point (miles)

R_η = nth zone radius (feet)

λ = wavelength (cm)

(b) *Refractive bending*

Refraction is the bending of the radio wave due to changes in the dielectric constant of the troposphere, which can be allowed for by using the modified refractive index, or alternatively, a modified earth radius. The refractive property of the beam is usually denoted by a k profile, where k is defined as

$$k = \frac{\text{effective earth radius}}{\text{true earth radius}} \tag{5.12}$$

The vertical distance between flat earth ($k = \infty$) and the effective earth at any given point can be calculated from

$$h = \frac{d_1\, d_2}{1{\cdot}5k} \tag{5.13}$$

where d_1 and d_2 are the distances in miles from the given point to each end of the path and h is the vertical distance in feet. This formula permits the plotting of an effective beam profile for a given k so that the earth may be considered a straight line.

The choice of k depends very much on the local weather conditions in the vicinity of the link, together with clearance on the degree of reliability to be ascribed to the link. For this reason, average values of k are chosen which relate to mean weather conditions over an area, in the case of the British Isles $k = 0{\cdot}7$ is used (over the American continent, where more varied conditions are to be found, various values of k are used from 0·3 to 1·33). Also the choice of k and clearance is linked to a propagation reliability which for United Kingdom links is 99·99 per cent (i.e. 0·01 per cent outage time or 53 minutes per year).

Fading

Fading occurs on all radio paths and can usually be broken down into two components, one of long period and the other of relatively short duration.

Long period fades occur due to diurnal variations in the dielectric make up of the troposphere, due to temperature, humidity and pressure changes. Transmission is best at mid-day and becomes progressively worse towards night time. Fading is most pronounced just after sun-down and sun-rise, more so in summer than in winter due to the higher water content causing beam bending.

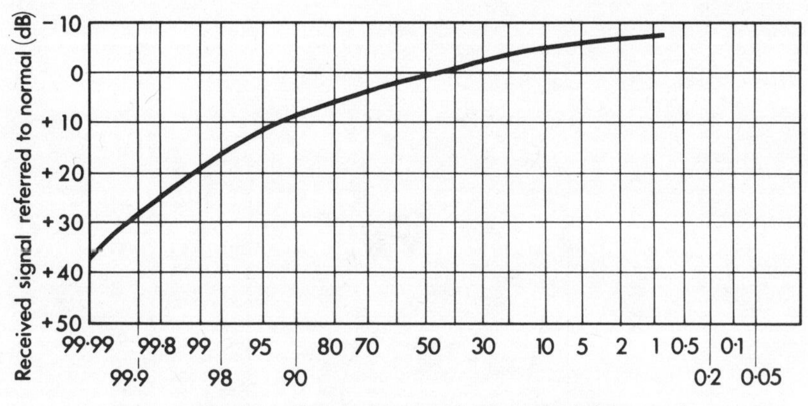

Figure 5.15 Fade margin based on Rayleigh scattering

As the length of the path is increased, the reflected waves from many indirect paths, including ground and atmospheric reflections, also increase, causing a greater random variation around a median level, as shown in Figure 5.15 (although this level is actually the r.m.s. value of the variation, it is usually permissible to associate it with the free space value). This will be referred to later in connection with link planning.

Absorption fades due to precipitation are not normally troublesome below 10 GHz, but when a wavelength becomes comparable to the rain drop size, this form of extra attenuation needs to be accounted for.

Generally speaking, fading can be compensated by increasing gain or transmitted power (for multipath types anyway) but this has its drawbacks and limitations. It is far more usual on troublesome paths to employ diversity techniques. Frequency diversity relies on the fact that two separate frequencies will fade by different amounts. With signals separated by greater than about 100 MHz, fading is incoherent and diversity combiners may be used to select the best output from several channels. Space diversity on the other hand uses two or more receivers, separated by a few wavelengths (usually vertically on the same mast). The former compensates for atmospheric fading and the latter for multipath.

Propagation path planning

The following is a summary of the manner in which paths are planned: (for a more complete summary, see references [19] and [20]).

(a) Select the most direct route using ordnance survey maps, path surveys, local authority planning searches. Avoid over-water paths if possible.

(b) Break up into sections (approximately 25-30 miles (or 40-48 km)) placing repeaters, so as to make maximum use of high points such as hills and tall buildings, and with due consideration to access.

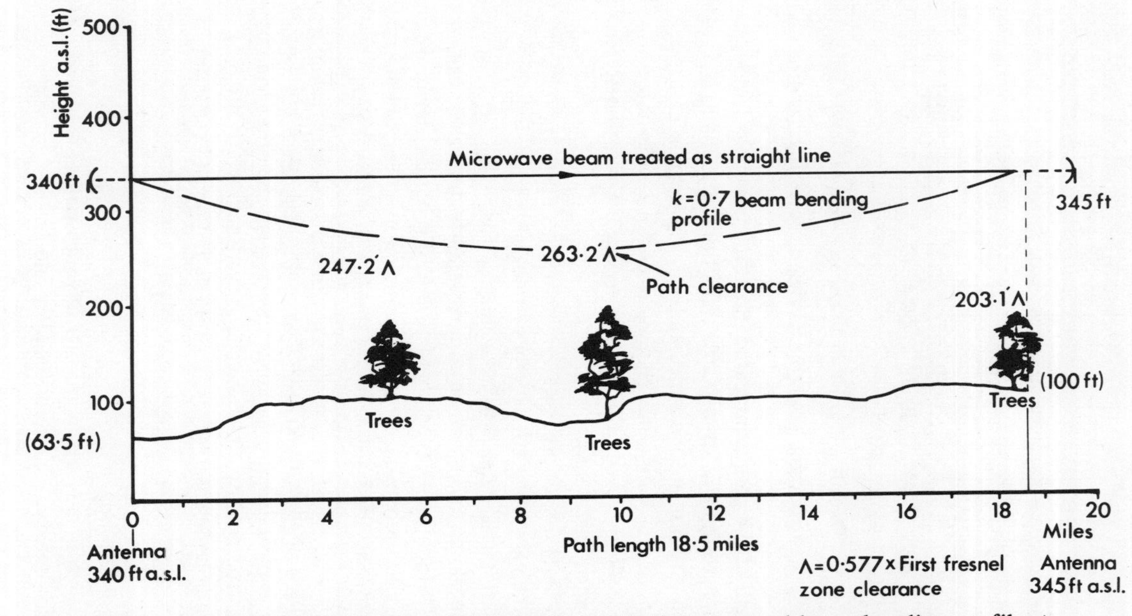

Figure 5.16 A typical link planning chart showing path clearance and beam bending profiles to determine antenna heights

(c) For each link draw a path profile (as shown in Figure 5.16) marking major obstructions and path clearances above gradient level. Draw profile for given k (calculated from given weather data and reliability) and superimpose on to path profile so that it just touches the clearance (pecked) points. Read off heights of masts. These may be adjusted to fit standard sizes or include vantage points.

(d) Compute expected fade margin from reliability data.

5.7 Tropospheric scattering

Tropospheric scatter communications may be defined as a means of transmitting microwaves in the u.h.f. and s.h.f. bands (900 MHz-5 GHz) via the troposphere, between points on the earth's surface of from 100 to 1000 km apart. Tandem operation of spans can yield communication circuits some thousands of kilometres in length. To date tropospheric scatter is the only practical wide band reliable ground based method of achieving long-hop communication. The advantages of the system are in its wide bandwidth, yielding multi-channel networks and high traffic density, whilst its disadvantage is in the use of high power transmitters and sensitive receiving equipment. However, it remains the only method of spanning large areas of inhospitable country with a reliable multi-channel communication link.

Although a large number of publications exist on the subject, as yet there is still no satisfactory theory for over-the-horizon propagation which fits all the observations and experimental data. There is no doubt that perturbations in the atmosphere is the cause of deviations from the expected diffraction field, which assumes empty space, but the exact mechanism by which energy is transported is still not known.

Originally it was assumed that scattering occurs within the common volume formed by the intersection of the antenna beams, as shown in Figure 5.17[21]. The theory was based upon the forward scattering of energy from turbulent eddies created by circulating wind conditions, and although it gave its name to the general form of communications, field strength predictions were lower than those actually measured. More recently, other theories have been forwarded, one of which ascribes the phenomena to a stratified atmosphere consisting of layers of differing dimensions at varying heights, producing reflection and refraction components at each layer[22]. Another theory[23] associates the phenomena with thermals (pockets of warm air created by circulating winds) and their lens like action in diverging the radio beams.

Clearly no concise theory exists, and for practical planning purposes use is made of empirical data obtained from exhaustive experimentation. Path attenuation is found to increase above free space (taking the angular distance travelled by the beam) by an over-the-horizon factor dependent on distance and frequency. The latter also depends on the temperature and humidity of the troposphere and explains why in humid tropical regions, e.g. the Caribbean islands, the extra

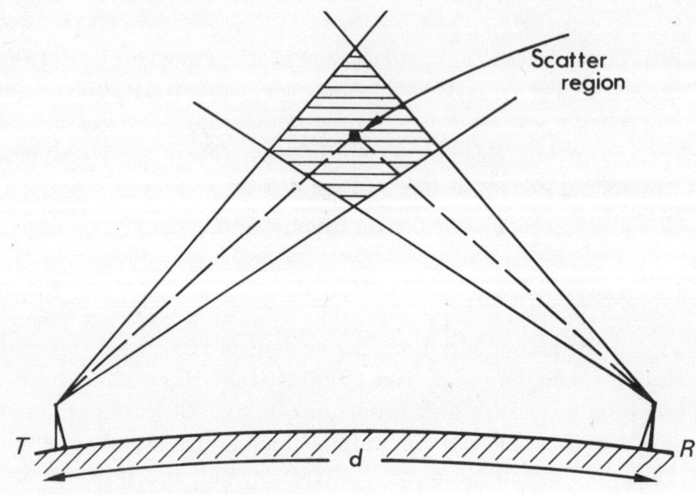

Figure 5.17 Tropospheric scattering

loss is about 10 dB less than in temperate zones and why in far northern regions a loss of 5 dB more is encountered. This consequently gives rise to a temperate zone variation which shows a maximum during the cold months (worst month February) of the winter and a minimum during the summer.

Both long term and short term (Rayleigh type) fading is encountered on tropospheric links and it is usual to employ a combination of space and frequency diversity called quadruple diversity to compensate for the more drastic variations in signal strength encountered. The quite severe multipath propagation associated with troposcatter effectively limits the usable bandwidth which can be transmitted without impairment of the modulated signal. Time differences of between 1 ms and 0·1 ms (depending on the length) are typical on links, and thus there is a trade-off between bandwidth and distance. The CCIR have laid down acceptable requirements for tropospheric links which are embodied in a very comprehensive design manual issued by the National Bureau of Standards[24], but a typical link operating on 99·9 per cent reliability for 200-300 channels covers path lengths of 160-240 kilometres. An excellent review article of the state-of-the-art due to Gunther[25] includes a complete list of all the systems in operation to date.

5.8 Extra-terrestrial propagation

All communications between the earth and outer space must pass through the troposphere and ionosphere, but outside the earth's atmosphere the effect on radio waves is negligible.

Essentially, above ionospheric critical frequencies, and certainly in excess of 100 MHz, the ionosphere is transparent to radio waves, although multipath propagation can result due to magneto-ionic splitting of the wave into ordinary and extraordinary components which can be accompanied by slight polarisation rotations. However, these effects become negligible at GHz frequencies and the ionosphere can be ignored.

The troposphere on the other hand contains weather and is frequency selective, allowing some frequencies to pass through readily whilst severely attenuating others. A range of frequencies in which waves readily penetrate is called a 'window'. The main window is that which ranges between ionospheric critical frequencies and frequencies strongly absorbed by precipitation (100 MHz-20 GHz). Other smaller windows can be seen from Figure 5.18

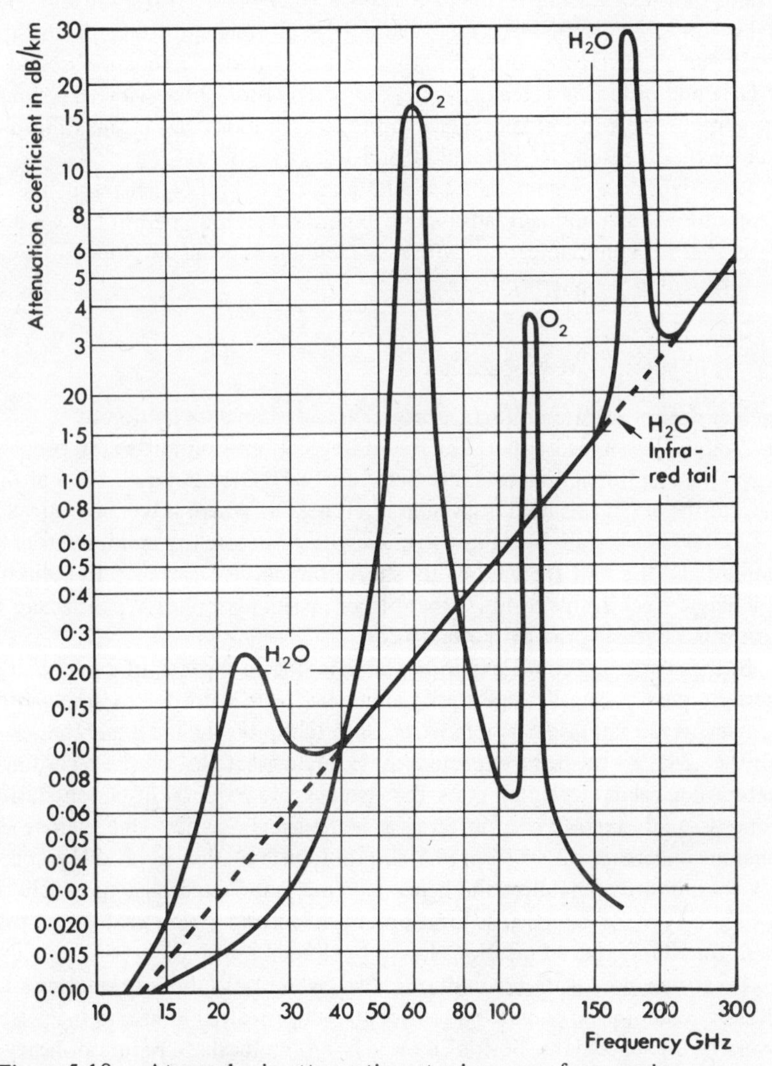

Figure 5.18 Atmospheric attenuation at microwave frequencies—a water vapour content of 10 g/m^3 is assumed. (From J. H. Van Vleck, *Phys. Rev.* 1947, **71**, pp. 413 and 425 and G. E. Becker and S. H. Autler, *Phys. Rev.* 1946, **70**, p. 300)

centred around 35, 90, 140 and 240 GHz, becoming increasingly narrower and more highly attenuated. Beyond this is the infra-red window from 1-1000 THz.

To date the lower GHz range between 3 and 7 GHz has been used for earth-space communications because this represents a minimum in both the attenuation and noise spectrums. However, in an already overcrowded spectrum below 10 GHz, space and terrestrial systems share common bands and this has produced problems of interference between them which has necessitated constraints on energy radiated from space and on the directivity of systems. Attention is now turning to the bands above 10 GHz for both space and terrestrial systems. As can be seen from Figure 5.18, although atmospheric absorption is fairly small, water vapour in the troposphere exhibits a resonance around 22 GHz and produces a steadily increasing attenuation above 10 GHz. A much more intense absorption takes place in the oxygen molecules in the atmosphere which exhibits resonances around 60 and 120 GHz.

Although these characteristics would tend to prohibit higher frequency communication equipment, advances will almost certainly permit the windows to be used for communications with new techniques being developed to combat the different atmospheric effects.

5.9 Millimetre wave propagation

A great deal of research effort has been expended, in anticipation of new band allocations by the 1971 Radio Regulations Committee, on propagation effects in the millimetre range between 11-50 GHz. An indication of the state-of-the-art in this field is given in reference 26, where it will be noticed that the 11-18 GHz band is being investigated with both satellite and terrestrial links in mind and the 35 GHz window for short-hop (about 5 km spaced) radio links. The main effect in this region is that of rain, particularly heavy storms and snow storms in increasing the path attenuation.

In the absence of much experimental data due to the lack of equipment in these frequency ranges, a crash programme has been instituted to supply attenuation versus rainfall rate characteristics, both theoretically and experimentally[27,28]. As might be expected, these rely a great deal on the recording of meteorological data; rainfall rates, for instance, are required instantaneously along the path, instead of at intervals as is the usual practice. One then needs reliable conversion curves from rainfall rate to attenuation to be able to plan a link for a desired reliability and hence fade margin. Such design curves are shown in Figure 5.19 for a particular location but the extrapolation problem of using them for widely spaced locations has not yet been resolved.

These new propagation conditions have given rise to the introduction of new diversity techniques; parallel path diversity for terrestrial links and space station diversity for space links. Both of these rely on the localised nature of heavy rainfall and the diversity advantage to be obtained between routes which are only spaced a short distance apart (a few kilometres)[29].

To obtain higher capacities, frequency re-use will almost certainly be

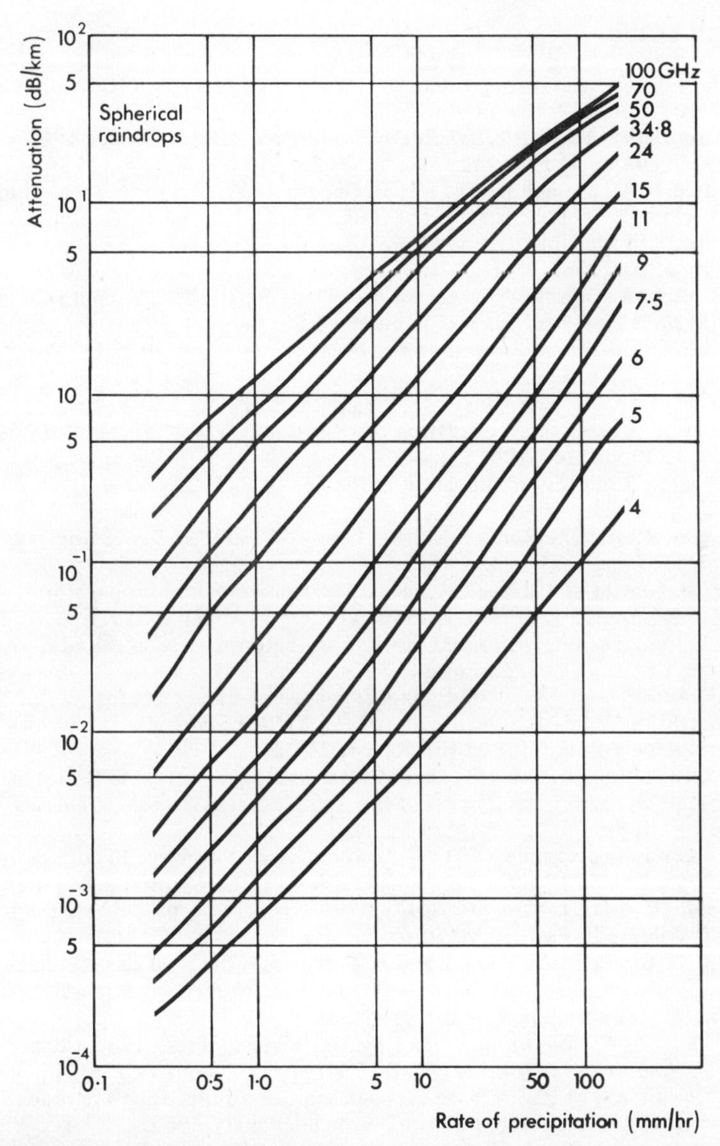

Figure 5.19 Theoretical relation between the attenuation and the rate of precipitation assuming spherical raindrops (from ref. 27)

employed, utilising the same frequencies on two orthogonal polarisations. This introduces problems of precipitation effects on cross-polarisation discrimination which are currently being studied. The majority of future radio links are also likely to carry digital rather than analogue information, and these are likely to be subject to impairments of transit time variations[30], and of fine-grain high-speed fading which could both cause excessive errror rates and limit link performances.

REFERENCES

1. *Radio Communications Regulations*, I.T.U., Geneva 1959. (To be updated by World Administrative Radio Conference, Geneva 1971).
2. *Extraordinary Administrative Radio Conference*, I.T.U., Geneva 1963.
3. *CCIR Documents of the IXth Plenary Assembly*, Los Angeles (1959) Volume 1, Recommendations I.T.U., Geneva 1959. (To be updated as in 1).
4. K. A. Norton, 'The Propagation of Radio Waves over the Surface of the Earth and in the Upper Atmosphere', Part II, September 1937, *Proc. I.R.E.*, **25**, pp. 1203-1236.
5. C. R. Burrows, 'Radio Propagation over Plane Earth-Field Strength Curves', *Bell System Technical Journal*, January 1947, **16**, pp. 45-75.
6. K. Bullington, 'Radio Propagation at Frequencies above 30 MHz', *Proc. I.R.E.*, 1947, pp. 1122-1136.
7. A. Watt, *V.l.f. Radio Engineering*, Pergamon Press, pp. 180-194.
8. J. R. Wait, 'Terrestrial Propagation of v.l.f. Radio Waves', *J. Res. NBS* **64D**, No. 2, 1960, pp. 153-204.
9. R. Moore, 'Radio Communication in the Sea', *I.E.E.E. Spectrum*, November 1967, pp. 42-51.
10. T. Greenwood, 'The Radio Spectrum below 550 kHz', *I.E.E.E. Spectrum*, March 1967, pp. 121-123.
11. E. V. D. Glazier and H. R. L. Lamont, 'Transmission and Propagation', *H.M.S.O. Services Text Books*, 1958, Volume 5, Appendix 14.3.
12. *N.B.S. Ionospheric Prediction Charts*, Washington, U.S.A. and Radio Research Station, Slough, England.
13. CCIR Report 248, 'Availability and Exchange of Basic Data for Radio Propagation Forecasts', *Documents of the XIIth Plenary Assembly, New Delhi 1970*, Volume II, Part (ii), Report 248-2.
14. CCIR Recommendation 399, 'Bandwidths and Signal-to-Noise Ratios in Complete Systems', *Documents of the Xth Plenary Assembly, Geneva 1963*, Volume III, pp. 24-26.
15. CCIR Study Programme 198 (VI), 'Estimation of Sky Wave Field Strengths and Transmission Loss for Frequencies Between the Approximate Limits 1·5 and 40 MHz', *Documents of the XIIth Plenary Assembly, New Delhi 1970*, Volume II, Part (ii), p. 70.
16. CCIR Recommendation 371-1 and Report 246, 'Choice of Basic Indices for Ionospheric Propagation', *Documents of the XIIth Plenary Assembly, New Delhi 1970*, Volume II, Part (ii), pp. 19-30.
17. J. A. Betts, *H.F. Communications*, English University Press, 1967, Chapter 10.
18. *Telecommunications Journal* (April 1964, pp. 119, January 1966, pp. 43-47), 'Prediction of Radio Wave Propagation Conditions using the Index I_{F2}', Bulletin A, Radio Research Station, Slough, January 1963.
19. K. Dumas and L. Sands, *Microwave System Planning*, Hayden, 1967.
20. H. Carl, *Radio Relay Systems*, Macdonald, 1966.
21. H. Brooker and W. Gordon, 'A Theory of Radio Scattering in the Troposphere', *Proc. I.R.E.*,* April 1950, pp. 401-412.
22. H. Friis, A. Crawford and D. Hogg, 'A Reflection Theory for Propagation Beyond the Horizon', *B.S.T.J.*, **36**, May 1957, pp. 627-644.
23. F. Belatine, 'Inadequacy of Scatter Mechanisms in Tropospheric Radio Propagation', *Nature*, **184**, November 1959, pp. 558-559.
24. P. Rice, A. Longley, K. Norton and A. Bursis, 'Transmission Loss Predictions for Tropospheric Communication Circuits', *N.B.S. Technical Note 101*, Volumes 1 and 2, May 1965.

25. F. Gunther, 'Tropospheric Scatter Communications–Past, Present and Future', *I.E.E.E. Spectrum,* 3, No. 9, September 1966, pp. 79-101.
26. *I.E.E.E. Transactions on Antennas and Propagation*, Volume AP-18, No. 4, Special issue of millimetre wave propagation, July 1970.
27. T. Oguchi, 'Attenuation of Electromagnetic Waves due to Rain with Distorted Raindrops', Parts 1 and 2, *Journal of Radio Research Laboratory, Japan,* 7, Sept. 1960, pp. 467-485 and Vol. II, Jan 1964, pp. 19-64, also Vol. 13, 1966, pp. 141-157.
28. R. G. Medhurst, 'Rainfall Attenuation of Centimetre Waves: Comparison of Theory and Measurement', *I.E.E. Transactions on Antennas and Propagation*, AP-13, July 1965, pp. 550-564.
29. R. Wilson, 'A Three-Radiometer Path-Diversity Experiment', *B.S.T.J.* July/August 1970, pp. 1239-1242.
    A Penzias, *'First Results from 15·3 GHz Earth-Space Propagation Study'*, *B.S.T.J.* July/August 1970, pp. 1242-1245.
30. D. Gray, 'Transit-Time Variation in Line-of-Sight Tropospheric Propagation Paths', *B.S.T.J.* July/August 1970, pp. 1059-1068.

Chapter 6

Radio systems

6.1 Introduction

The previous chapter has dealt with the general characteristics of radio propagation. This one deals with the problems of using radio transmission in a system, and in particular the use of frequencies below 1 GHz with little directivity. The main uses of radio are:

(a) broadcasting,
(b) point to point over long or inhospitable distances,
(c) mobile operations to ships, planes, cars and policemen.

The general advantages of radio are obvious, in that one can have broadcast and mobile operations, and the major disadvantages of radio are the dependence upon the physical environment and also, nowadays, the popularity of the method, which means that the spectrum is very crowded giving only limited usage. Although there are improved techniques for utilising the bandwidth more efficiently and for using a greater range of frequencies, the bands available are definitely limited and it will not be long before only essential traffic can be carried by radio.

The most straightforward application is to broadcasting, and here the problems are mainly ones of frequency allocation for different stations carrying the same programme. The problem resembles a huge jigsaw puzzle. Prior to 1947 there was little or no co-ordination of organization and the ITU mainly acted as a register of frequencies used. However, in 1947 the Atlantic City Conference agreed to international co-ordination. This committee shares out frequency blocks which the individual administrations, i.e. the Ministry of Posts and Telecommunications here and the Federal Communications Commission in the United States, share out in detail, and this is far from limitless.

Point-to-point systems operate either in the h.f. (3-30 MHz) band for long distance use, or in the v.h.f. and u.h.f. band for line-of-sight working. This chapter concentrates on the h.f. services since it is these which introduce the main system problems when they are to be integrated into the public telephony system.

6.2 H.f. Radio systems

H.f. radio has been the major point-to-point contact with other countries apart from telegraph cables. The first commercial radio telephone service between the United Kingdom and the United States was operated in 1927 and this grew to about ten to twelve audio circuits only by 1956 when it was supplemented by the submarine cable. It is still widely used for communication with countries which are not yet connected by cable or satellite, and obviously is retained as a standby for any cable breakdown. Otherwise it is used mainly for telegraph operations.

There was little demand for service on the public network, mainly because of cost and unreliability, and the public network of radio links has failed to expand as spectacularly as the other communication fields. There is, however, a very strong demand for private services which terminate in a customer's office. This again is mainly for telex traffic since there is little hope of finding many more 3 kHz spaces in the spectrum.

The technique of integrating h.f. radio into the public telephone system is involved. The method which has emerged is based on single-side-band amplitude-modulation, primarily for the bandwidth saving and the better signal-to-noise ratio that may be obtained compared to double-side-band a.m. The use of s.s.b. in the past has meant that some form of automatic frequency control (a.f.c.) has been necessary, but with modern equipment this is no longer necessary since it is possible to generate transmitter and receiver carriers with an adequate degree of stability [1].

The original technique was to transmit 2 x 3 kHz telephonic signals with a 3 kHz gap between them to avoid third order intermodulation products. This is called i.s.b. (independent side-band). This would then be transmitted with a reinserted carrier which could then be extracted at the far end and used for automatic frequency control. However, since the late forties the linearity of the transmitters has been improved and some multichannel telephony is possible. The usual technique is 4 x 3 kHz channel in a 12 kHz band with a low level carrier. These channels may then be combined as a programme circuit or split into multi-channel telegraphy.

The design of h.f. transmitter stations has changed almost completely over the last few years [2]. The propagation of these frequencies is a hazardous business which is at the mercy of the time of day and sun-spot cycles, etc. To operate such systems it is necessary to change the frequency during the day from that during the night and at other times depending upon the recipient country's condition. This has needed a considerable amount of man-power to perform these operations and keep watch over the circuits, but nowadays the system is increasingly being automated [3].

The secret to this development is the self-tuning transmitter and receiver. The basis of this method is to produce a frequency synthesiser working from a high grade frequency standard (1 in 10^8 typically). With such devices it is possible to synthesise a frequency by means of setting it on a decade dial and these frequencies may be used as the input to several systems with automatic phase control which tune the capacitors of the transmitters and receivers. In addition systems are incorporated which adjust the coupling so as to produce maximum loading to the antenna. A modern transmitter can change frequency automatically in 10-15 seconds.

Groups of transmitters and aerials are then incorporated in a motor driven switching array so that any transmitter may be associated with any aerial. This technique also allows the economic use of *dualling* whereby a second transmitter is used on the new frequency while the main one is changing, and so it provides continuous service.

Modern prediction techniques are so good that these frequency changes can be pre-programmed.

Integration in the public telephone system

Because of the high-power transmitters needed, and the high sensitivity receivers, it is desirable to locate them long distances apart. They are then controlled from a central position, which in the United Kingdom is the Radio Telephone Terminal on the outskirts of London. Speech signals are normally received at the control point on a 4-wire basis and they will arrive with a wide range of power levels. Since a transmitter which would give acceptable signal-to-noise ratios at the lowest level, but not distort at the highest, would be totally uneconomic, then it is necessary to have some automatic gain amplifier which can give a constant output volume to ensure that the transmitter is being loaded efficiently. These *constant volume amplifiers* (c.v.a.'s) are arranged to have a fast response time in the order of a few milliseconds, and then have a hang-over time of the order of seconds. Typically these amplifiers will reduce the dynamic range by 15 dB.

When the radio signals are transmitted via the ionosphere, they are subject to general and selective fading. Whereas in d.s.b. the average level of the received signal may be used for automatic gain control, s.s.b. requires the use of a pilot and because this may be subject to selective fading, then a long time-constant (of the order of 10 seconds) is needed for the a.g.c. circuit. Although the output of the receiver is maintained at substantially the same average level, it may still be subject to short term variation of the order of 20 dB. The variations have to be corrected by another constant volume amplifier. In addition to constant volume amplifiers and a.g.c., compandors and expandors are also used to reduce the background noise. It is important to realise why compandors, constant volume amplifiers and a.g.c. are all used in the same system.

Since the output volume from the receive c.v.a. is effectively constant whatever the input volume to the transmit c.v.a., there is a highly variable gain through the system which depends upon the speech level. If the system is connected to the public network at both ends, then it will meet the 2-4 wire hybrids at the International exchanges, and because of the low loss across the hybrid there is a high probability of instability (or singing). It is necessary, therefore, to fit a singing suppressor which is effectively a voice operated relay which makes the circuit into half duplex working.

On many systems privacy equipment is also provided which scrambles the speech by splitting the signal with several narrow frequency bands, and then re-arranges these bands so as to make the speech unintelligible unless an un-scrambler is used. There are usually switched by the singing suppressors, since this is a large and expensive piece of equipment. A complete block diagram of the h.f. telephone system is shown in Figure 6.1.

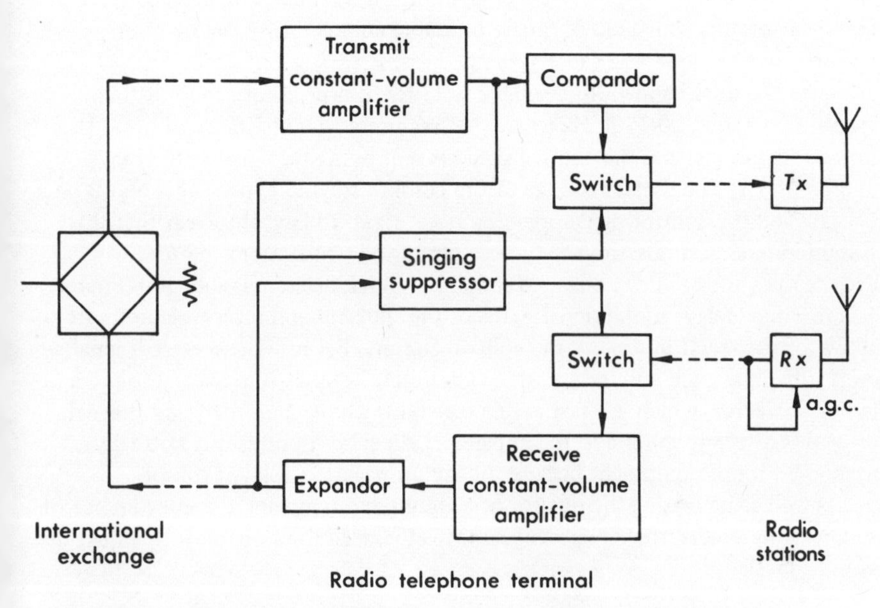

Figure 6.1 Simplified diagram of h.f. radio telephone systems
Basic characteristics
(1) **Transmit c.v.a.** Input –50 to –10 dBmO, Output –10 dBmO ± 3 dB. Reaction time: Increasing signal –2 ms with correct value attained after 20 ms. Decreasing signal: Unchanged for 1 s, correct value attained after 5 s.
(2) **Receive c.v.a.** Input –30 to –10 dBmO. Output –10 dBmO ± 2 dB
(3) **Compandor** 2:1 law in dB. Syllablic reaction time of 20 ms
(4) **A.g.c.** From pilot with 10 s time constant

Note that the operation requires the attention of two operators at each end, one for traffic and one for technical control, although these share their services among several lines. It is necessary to make many changes in the operating frequencies, and this is usually done by the use of tunable transmitters.

There is as yet no automatic working of these lines, although telex calls can be dialed through some radio links.

Lincompex system

A modern development which has revolutionised h.f. transmission and produces circuits which are fully competitive and interchangeable with cable systems is called Lincompex [4] (LINked COMPression and EXpandor, a British invention). The basic idea of Lincompex is simply to transmit a control signal which gives a measure of the amount of volume compression at the sending end. At the receiver this control signal then adjusts a complementary volume expandor so that the overall gain of the circuit may be held constant irrespective of the changes of path attenuation. The effect will be to change the level of background noise, and since it is now possible to use faster acting (20 ms) expandors and compressors, then these give a companding advantage. The overall effect is that even circuits which are normally unusable commercially can be given performances as good as a cable system.

Since the total bandwidth available is 3 kHz, it is necessary to limit the speech to a band 300-2700 Hz (the CCIR recommend a minimum of 2600 Hz for a radio circuit) and the control signal is sent by a f.m. signal in the range 2900 ± 90 Hz. The dynamic range of the compressor and expandor is 60 dB on a 2:1 law and the control signal operates at a 2 Hz/1 dB law. It is clear that this system implies accurate frequency tracking of the two stations in order to decode the control information. A simplified block diagram is shown in Figure 6.2; the time delays in the signal paths of the transmit and receive sections are to allow for the reaction time of the control circuits. Because the received signal, even with a.g.c., will still have a wide variation of level, it is necessary to use a fading corrector, which is a fast acting constant volume amplifier, and this will present a constant volume to the expandor. An echo suppressor is normally needed with this equipment because of the propagation delay involved.

As well as increasing the quality of a circuit and removing the interference of singing suppressors, the equipment removes the need for a operator to adjust the volume levels.

6.3 Mobile operations

The other major application of radio is for mobile operation and this has been a post-war development. In 1945 there were less than 1000 civil radio-telephone systems in the United Kingdom and the majority of these were for fire and civil defence. There was no ambulance, taxi, marshalling yard or commercial concern with any radio equipment. Today this has radically changed, and there are upwards of 70 000 in Britain, mainly fitted to cars in various fleets. The main reason for this has been the evolution of a small reliable transceiver of small bandwidth. These usually operate in the 80 MHz, 180 MHz, and more recently, the 450 MHz bandwidth. In the United States experimental work is proceeding in the 850 Hz band. The higher the frequency then the smaller are the antennas that are needed, but the practical range is less. Also there is more spectrum available in the higher bands, but to utilise it effectively requires very high stability

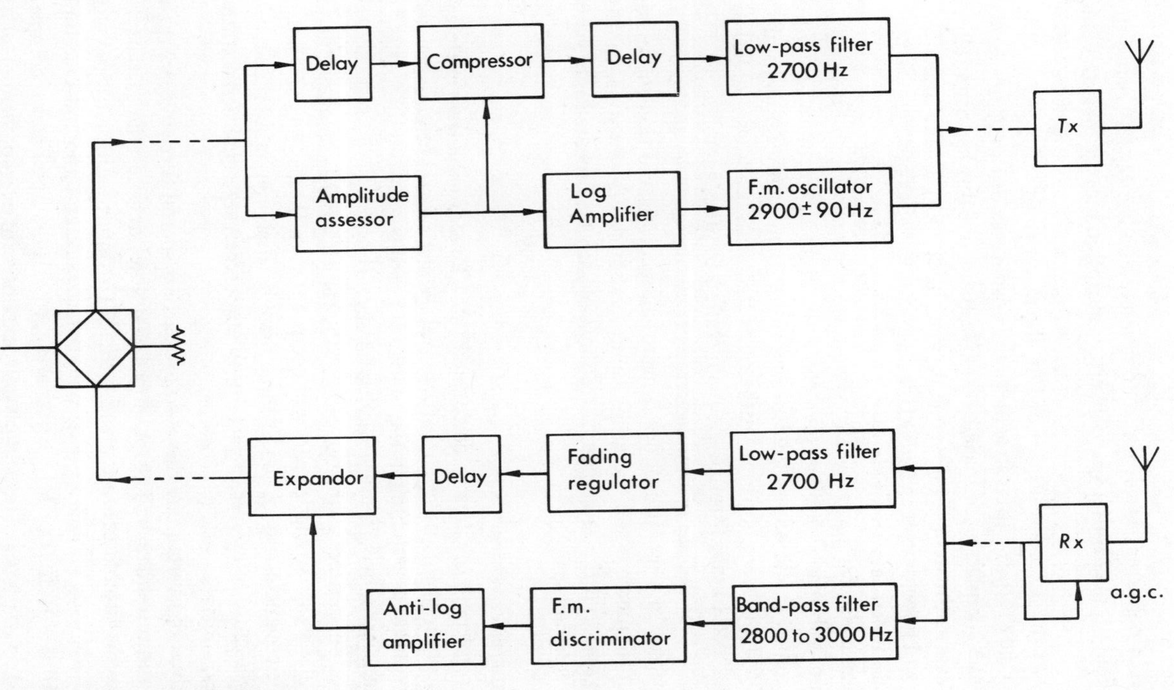

Figure 6.2 Lincomplex system

oscillators e.g. ±10 Hz stability at 900 MHz needs oscillators of about 1 in 10^8 stability. This sort of stability is necessary for vehicular systems since at these higher frequencies there is a significant Doppler shift which will modulate the received signal.

The use of high frequencies is not completely disadvantageous, since the short wavelength means that space diversity reception may be used to combat fading.

In spite of the noise advantages of f.m. the majority of sets use a.m., since the increased bandwidth required to achieve the advantages of f.m. is not available. Originally the channels were spaced 100 kHz apart, but this has been reduced over the years as the demand has gone up and the stability of the oscillators increased, and nowadays it is 12·5 kHz.

The majority of sets are of simplex systems on double frequency working. The disadvantages of a 'press-to-talk' system are not so great, and if there is one main station with many mobiles, then the short procedure achieved by 'press-to-talk' is desirable. It also allows loudspeaker rather than handset operation. Duplex working is used, however, at the main station as it allows mobiles to break-in and also to talk through the main station to other mobiles.

It is found that the use of double frequency working increases the useful working space of the spectrum since it means all receivers and transmitters are on different frequencies and therefore do not readily interfere with each other. This means also that the same pair of frequencies may be allocated more frequently geographically.

Problems of area coverage for mobiles

Problems arise where the area to be covered is too large for the transmitter. One major application of this is for aircraft control. Whilst separate frequency allocation with band switching in the mobile is possible, this is undesirable, especially in a fast moving vehicle such as an aircraft.

With two transmitters giving out a.m. at the same carrier frequency, then in areas of similar signal strength beats will occur between the carriers. Even if the transmitter frequencies are locked in some manner, then inevitably phase variations will cause interference patterns in areas receiving equal signal strength from two transmitters.

One method of overcoming this is to use carriers spaced further apart than the highest audio frequency. This produces beat notes between the carriers, but the beats may be filtered off in the audio stage. The input radio frequency stages have to be made wide enough to receive all the carriers, and no band-switching is necessary. Thus with four offset carriers arranged in a diamond mesh, complete coverage may be achieved. This is, however, expensive in overall channel width and makes the noise bandwidth of the receiver higher. With very high stability transmitters it is possible to make the differences in frequency very small and produce sub-audio beat notes.

Another problem is that the same signal must be radiated for each transmitter, and if the audio is sent over a carrier system or a landline, then significant differences in time delay may occur (up to 15 ms typically); also, if part of the route is on a carrier system then frequency transposition may occur. This problem is even greater on the receiver side since all receivers are waiting for one signal, which may come from anywhere, and all will be contributing noise, although a threshold muting circuit is possible. Also the time delay problem is oppressive and the end product is intelligible, but not good quality speech. So far no passive network or magnetic-drum solution has proved satisfactory in equalising land line circuitry. The only solution is extensive use of radio links. Meanwhile, evening-up of line lengths is possible but complicated by the fact that it is necessary to provide alternative routing to all stations for security reasons. The need for area coverage is increasing as air traffic control is becoming more sophisticated. Also 50 kHz channels are protected until 1972, but demand will reduce spacing to 25 kHz thereafter.

Research is proceeding into ways of producing area coverage within these narrower bands. If the main stations are synchronised then in areas of approximately equal strength the receiver input will vary with phase drift. If now the receiver is moved rapidly across this field then violent chopping will occur which will make the signal unrealiable. Experimentally, it is found that if the signal strengths differ by 15 dB then the effect is more or less unnoticeable. Therefore in areas of signal strength differing by less than 15 dB, an off-set station is situated which has a strength of 15 dB above the combined other stations. This introduces very tricky siting problems.

Public mobile telephony

The requirements for a public mobile-telephony service are more stringent than those for private systems as outlined above. The major differences are that a much better quality of service is expected together with an operational procedure similar to a conventional telephone. From the technical point of view the problem is complicated further by the fact that signals arriving at the transmitter will have a wide range of speech volumes. In the United Kingdom a public radio-phone service is available in the London area in the frequency band around 160 MHz. This uses three traffic channels in each of three overlapping areas, together with a common calling channel. Although 25 kHz channelling is used nominally, careful siting of the transmitters has permitted the use of 50 kHz channelling at any particular station, and this allows the use of phase modulation with a deviation of ±10 kHz. This gives a 3 dB signal-to-noise advantage. The transmit and receive channels are fitted with constant volume amplifiers. Since there is no need for a hybrid transformer at the mobile station, there are no problems of oscillation.

This system is under operator control and uses selective calling on the common channel to alert the mobile. The mobile answers or initiates a call on an idle free channel selected by the user. The equipment is currently being updated by providing automatic idle channel selection in the mobile. All idle channels transmit a distinctive tone continuously and the receivers lock on to one such channel. The selective calling signal is sent simultaneously over all idle channels and hence there is no need for a separate control channel. Dialling facilities are provided in this equipment.

A similar system is in use in North America[5], but provides for fully automatic dialling to and from mobiles. The main disadvantages of the current system are that mobile equipment is still expensive and the use of spectrum is very limited. Higher frequencies are being used (up to 900 MHz) but the coverage problem is considerable for a complete city.[7]

Transmission to a fast moving vehicle such as a train presents further problems, even though the route is well defined. These may be overcome by the use of several fixed transceivers covering a particular route and making special provisions for tunnels etc. The provision of such a service is only justified at present on very heavily used routes such as Boston to New York in the United States of America. Coin box service is provided on some of these trains and a user is given a few minutes warning if a channel change is imminent as he moves from one service area to the next. He must then re-establish the connection if he wishes to continue.

Communication to aircraft is also a problem, as over large parts of the Atlantic only the highly variable h.f. region is usable. There are currently several proposals for using satellites for aircraft communication but as yet none of these systems have been implemented.

Paging systems

Because of the grave shortage of frequency spectrum, it is unlikely that any large scale system of public mobile telephony will emerge. However, an alternative system which will cater for most of the needs and that will not require much bandwidth is that of selective calling, or paging systems, such as are in operation in Switzerland and in Belgium and the Netherlands[6].

These are systems by which a subscriber has a small receiver which he can keep in his car. This may be activated by means of a dialled code on the public exchange. This code is received at a central point and a unique pattern of three audio frequencies selected from thirty is pulsed out; the receiver detects the pulses and lights a lamp. This gives a maximum of over 25 000 subscribers on any one carrier frequency.

In order to achieve nation wide coverage in Switzerland, these transmitters are necessarily on d.s.b., and to prevent interference their carriers are off-set by more than the highest audio frequency. The receivers then have a wide enough band-width to receive any one of the three stations. Radio links are used between the transmitter to avoid time delays.

The Netherlands have a more sophisticated method, where not only may a subscriber's set be activated but he may also be sent a coded message in the form of a digit 1 to 6. This, by some pre-arranged code, may indicate who wants him.

Again the system uses the public network and in fact the first electronic exchange for the Netherlands was installed for this purpose.

In the Netherlands system, area coverage is provided by having only one transmitter on one frequency at a time. Thus a signal is transmitted from the transmitter used in sequence. The codes have a very high capacity and the service is proving very attractive.

There are at present (1971) discussions which it is hoped will lead to an integrated paging system throughout Europe accessible from the telephone network. The proposed system will use four radio channels around 87 MHz and have a coding capacity of a million signals.

REFERENCES

1. For example, see J. A. Betts, *High Frequency Communications*, EUP, 1967.
2. W. J. Morcon, 'New Approach to h.f. Station Planning', *Proc. I.E.E.* **110**, 9, September 1963, pp. 1583-5.
3. Transmitter station design details may be found in a series of articles in the *Post Office Electrical Engineers Journal 59*, July 1966 to January 1967 issues. Receiver design is covered by E. G. Branson, 'The New Bearly Receiving Station', *Post Office Electrical Engineers Journal 61*, July 1968, pp. 75-81.
4. D. E. Watt-Canter and L. K. Wheeler, 'The Lincompex System for Protection of h.f. Radio-Telephone Circuits', *Post Office Electrical Engineers Journal 59*, 3, October 1966, pp. 163-7.
5. V. A. Douglas, 'The MJ Mobile Radio Telephone System', *Bell Lab. Record* **42**, 11, December 1964, pp. 382-9.
6. G. M. Uitermank, 'A New Service: A Country Wide Radio Code Paging System', *Philips Telecommunications Review* **24**, 1, February 1963, pp. 1-12.
7. E.g. see 'High capacity mobile communications system proposed' *Bell Laboratories Record,* **50**, March 1972, pp. 96-97.

Chapter 7

Microwave transmission systems

7.1 Introduction

A microwave radio system is similar to other carrier transmission systems such as cable and coaxial line already described, but with the conductor path between repeaters replaced by a radio (air) path. The design of such systems is therefore very similar to cable and wire circuits, and the choice between these alternatives depends on a variety of economic and practical factors.

Microwave radio systems operate in the frequency spectrum above 900 MHz, and comprise either the more conventional line-of-sight links, or over-the-horizon troposcatter links. The CCIR [1] have allocated the 2, 4, 6 and 11 GHz bands for such links, which have the advantage of wide bandwidths and small high directivity antennas. Coupled to this they have the general advantages of radio systems:

(a) Capital cost is generally lower.
(b) Installation is quicker and easier.
(c) Additional service may be provided quickly and cheaply.
(d) Irregular terrain difficulties are overcome.
(e) Equalisation need only be applied for the equipment as the frequency characteristics of the transmission path are essentially constant over the transmission bandwidths.
(f) Repeater spacing may be increased by increasing tower heights.

On the other hand, microwave systems have various disadvantages as compared with cables, including:

(i) The restriction to line-of-sight operation on conventional links.
(ii) The problem of suitable access to repeater stations from main highways and provision of accommodation for maintenance. A cable system would follow the road or railway.
(iii) The provision of power supplies for the repeater. This may be alleviated with an all solid state system which could run off its own generator (may be gas fuelled). At the moment United Kingdom repeaters on high traffic routes have stand-by generators or continuously running

alternators to provide power with no break. Cables may be provided with their own power supplies fed from a few locations.

(iv) It is difficult to provide short distance spur circuits to intermediate exchanges or subscribers.

(v) Adverse weather conditions can cause severe fading and beam bending which necessitates both frequency diversity operation and auxiliary switching to stand by channels.

(vi) The high level of linearity required in the repeaters poses a severe design problem.

Microwave links were being developed before the Second World War, and although war halted development to concentrate work on radar systems, it gave the klystron and travelling wave tube which played such an important part in future systems.

After the war it was television that forced the pace, and in 1949 the first radio-relay links operating in the 1 GHz range were being installed to transmit 405 line, 3 MHz television signals between the studios and the transmitters. It was originally planned to use a.m., but it was found that with f.m. the linearity requirements could be more easily achieved. The advantages of f.m. were thus realised at an early stage, and although television continued to be the dominant factor in the expansion of the network, it paved the way for the successful operation of multi-channel telephony links involving a number of telephone channels arranged in f.d.m. in the baseband spectrum which were to follow in 1956.

There has been a rapid expansion of microwave radio relay trunk networks in recent years[2] stimulated partly by the introduction of standard trunk dialing and partly by the requirements of the television authorities to provide increased numbers of television channels capable of taking colour programmes. The United Kingdom network now consists of trunk links between the major cities in Britain, a cross channel link and many spur links to smaller towns. The broadband channels are separated into those transmitting mult-channel telephony with capacity either 960 telephone channels (f.d.m. assembly of 4 kHz channels in baseband 60 kHz-4·028 MHz) or 1800 telephone channels (baseband 312 kHz-8·204 MHz) and video channels (baseband 0-5 MHz) for the transmission of 625-line television. This has led to the Post Office Tower[3] in central London and its 120 sisters scattered around the country. The Post Office Tower is located at the centre of the trunk telephone exchanges and the London network of television cables. It is the terminal for telephone and television circuits to the continent via cross channel link and to further parts of the world by way of the satellite ground station at Goonhilly and the network of satellites.

7.2 Microwave link equipment

The frequency allocation of microwave bands has been agreed internationally by the CCIR [1], and those used in the United Kingdom are shown in Table 7.1.

Table 7.1. Microwave band frequency allocations as agreed by CCIR

Band centre in GHz	Bandwidth in MHz	Channel spacing in MHz	Number of both-way channels
1·8	200	30	2
2·1	400	30	6
4·0	400	30	6
6·175	500	30	8
6·760	700	20	16
		40	8
11·190	1000	40	12

Links now operate in the 2, 4 and 6 GHz bands and research is progressing into the use of the 11 GHz and higher bands in which the attenuation due to rain complicates the path attenuation and link planning. Each band contains many individual broadband channels and the link planning feature of the propagated microwave beam between repeater stations has already been discussed in Chapter 5. Next typical microwave link equipment is reviewed as an example of the techniques generally applied.

(a) *Single broadband-channel link* One broadband (30 MHz) single channel link suitable for multi-channel telephony or television transmission is shown in Figure 7.1.

Transmitter

The baseband signal, after suitable pre-emphasis, modulates an oscillator with centre frequency 70 MHz (i.f., CCIR recommended). The peak-to-peak frequency deviation of the television signal is 8 MHz and the maximum (busy hour) r.m.s. deviation of multi-channel telephony is 1·1 MHz (for 960 channels). The modulator and demodulator must be highly linear to avoid intermodulation noise. The signals are amplified and converted to microwave frequency via the local oscillator/crystal mixer. The mixer output contains two f.m. side bands and, since each contains all the essential information, one of them is filtered out by the microwave filter. The signal is then applied to the travelling-wave-tube (level +1 dBm) of 36 dB gain. The output signal (level +37 dBm) is connected through a ferrite isolator and low-loss waveguide feeder to the antenna.

Echos in f.m. systems give rise to intermodulation noise in telephony systems and distortion in television systems which must be minimised. Echos arise from reflections from mismatched components, particularly of the antenna, and so isolators are inserted in various places to reduce these by at least 30-40 dB.

Transmission path

Features of this are dealt with in chapter 5. The usual repeater spacing is 25 miles (40 km), which at 6 GHz gives a free-space loss of 140 dB. Parabolic antennas having gains of about 40 dB with 2° beam width, are quite common

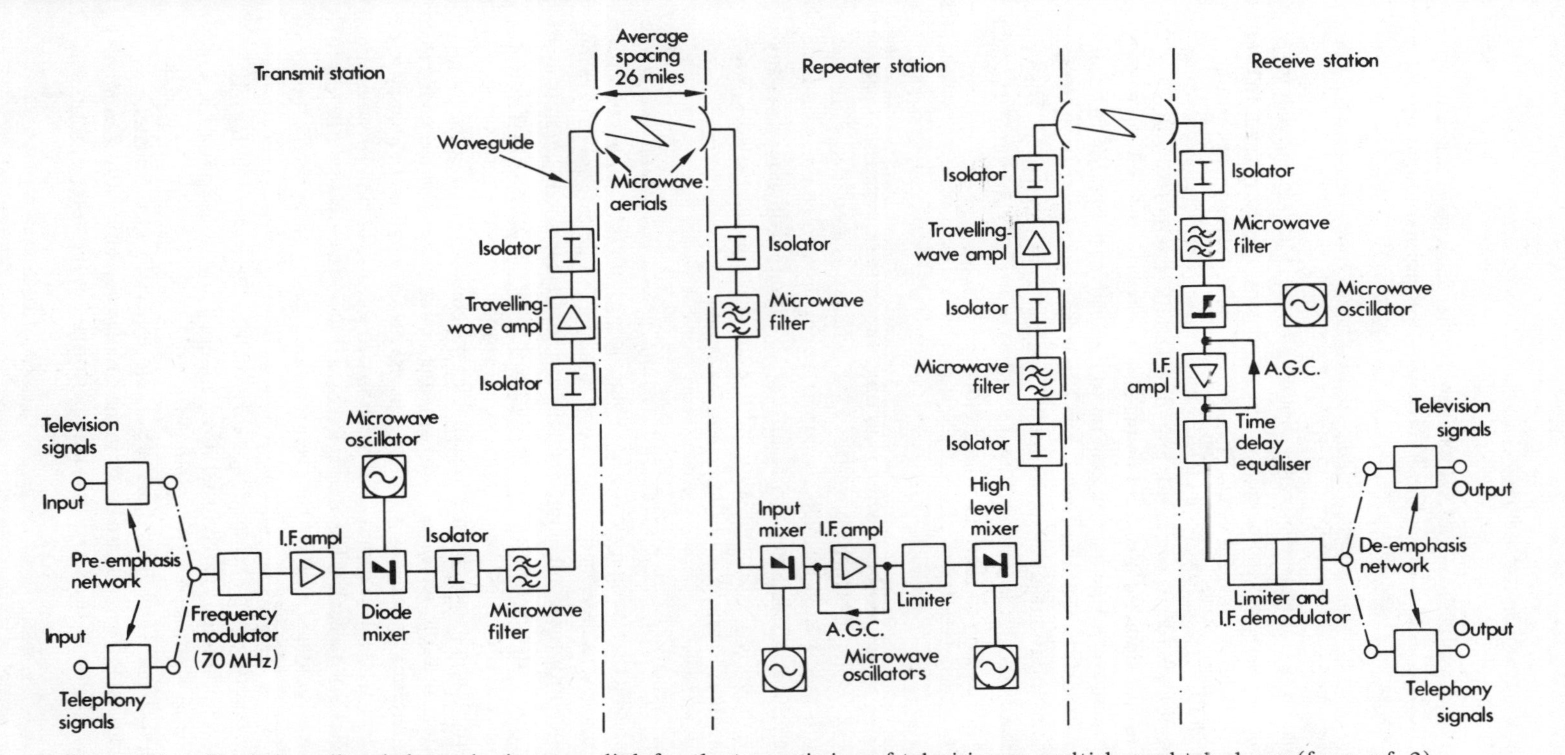

Figure 7.1 Single broadband-channel microwave link for the transmission of television or multichannel telephony (from ref. 2)

and their use results in an overall loss, including feeder losses, of about 67 dB, which is constant over the 30 MHz band for normal propagation conditions.

Repeater

The signal level at the input mixer is about –30 dBm, but this may fall to –60 dBm due to severe fades. Most of the fixed gain and reserve gain to overcome such fades is provided at i.f. The signal at the repeater is fed through an isolator and filter to a low noise (10 dB) input mixer/local oscillator, which convertes it back to i.f., where it is amplified (80 dB maximum) with a.g.c. control to give stablised output in fading conditions. The signal then passes to a limiter to remove any added a.m. and on to the high level mixer/local oscillator, which reconverts it back to r.f. with a 252 MHz frequency shift from the input. This enables input-output channel intermodulation to be minimised. The signal then passes through the isolator/filter (+1 dBm) to the r.f. travelling-wave-tube (36 dB gain) and on to the transmit antenna (+37 dBm).

Receiver

The received signal will be down to –30 dBm due to the transmission path at the mixer/local oscillator which converts it to i.f. where it is amplified with a.g.c. as before. It is then passed through a group delay equaliser to compensate for adverse equipment frequency characteristics, to a limiter which removes any residual a.m. on to the demodulator and relevant de-emphasis network, whose characteristic is the inverse of the pre-emphasis one.

(b) *Multi-channel broadband link* Multi-channel links utilise channel combining and separating filters (diplexers) to combine or separate the outputs of several single channels, as described in (a) to one antenna. Circulators are used extensively now, both for channel multiplexing and for higher power antenna multiplexing. These arrangements make use of the directional properties of ferrite circulators and the reflective or transmissive properties of the filters placed in their ports. Using parabolic antennas enables the connection of up to four transmitters and four receivers to the antenna in one band. The use of the horn reflector antenna enables the simultaneous use of two or three frequency bands (4, 6 and 11 GHz). It transmits and receives two signals via circular waveguide, polarised at 90° to each other, and is fed by a 3 inch low-loss (0·5 dB/100 ft. at 6 GHz) waveguide, separation between the modes being obtained by the use of mode filters to separate rectangular waveguide feeds. The horn reflector's gradually tapering feed enables a superior match to the waveguide feeders, reducing echos and mismatch losses. Conventional structures have a 8 x 4 m aperture with a gain of 45 dB. To minimise waveguide losses it is important to keep the rectangular guide as short as possible, especially at the higher frequencies, and to avoid sharp bends in the circular waveguide. Final combining and decombining of the signal is usually performed at the antenna from two separate rectangular waveguide feeds to a circular antenna feeder. In this way the only polarisation discrimination required is in the polarisation filters where 30 dB is adequate and easily achievable.

Channel frequency allocations

The 500 MHz bandwidth of the 6 GHz band is divided into two adjacent blocks 250 MHz wide, each block containing eight r.f. channels spaced 29·65 MHz apart; at any one station all transmit channels are allocated to one block and all receive channels to the other. To enable channel signals to be more easily separated, adjacent channels are orthogonally polarised. Channel assignments are shown in Figure 7.2, where all go and return transmitter channels ($f_1 - f_8$) are located in the low block and all receiver channels ($f_2' - f_8'$) in the high block. A frequency gap of 44·5 MHz is provided between transmit and receive blocks to reduce transmit/receive crosstalk, the guard band being 6·2 MHz wide.

In the arrangement shown in Figure 7.3, the individual broadband outputs of eight travelling-wave-tubes are combined into two groups (f_1, f_3, f_5, f_7 and f_2, f_4, f_6, f_8) at terminal station A by means of ferrite circulator channel combiners and connected to the antenna via a polarisation filter and circular waveguide feeder. An identical antenna and feeder system is used to receive eight broadband channels (f_1'-f_8') from the distant repeater system.

The repeater station is simply a terminal receiver and transmitter connected at i.f. with microwave local oscillators displaced in frequency to give the input-output carrier frequency shift of 252 MHz between high and low blocks (i.e. f_1, f_3, f_5, f_7 (H) into repeater come out as f_1', f_3', f_5', f_7' (H) etc.). The terminal station B is exactly the same as terminal A, providing reciprocal transmit/receive operations.

To achieve the reliability needed on a broadband trunk network, stand by or so called 'protection' channels are provided that automatically switch in to replace faulty working channels. Protection channels are shared between several working channels, usually one or two to every twelve operation ones, and switching is carried out at i.f. by high speed diode switches in the terminal stations. Each r.f. channel carries a continuity pilot signal located above the traffic signal frequencies in the baseband, should this fall by 6 dB or noise in a narrow band around the pilot increase beyond a predetermined limit automatic switching to a protection channel is carried out. Auxiliary narrow bands of the frequency spectrum are allocated for alarm from repeaters and terminals and for the transmission of signals associated with switching, control and supervision, A, B, C and D in Figure 7.3 (CCIR [1], Recommendation 389, 444).

Modern microwave radio equipment [4,5] employs solid state devices throughout to increase reliability and equipment stability and to reduce power consumption and overall equipment rack sizes. Tunnel diode amplifiers in the receivers are replacing conventional types due to their better noise performance. Crystal controlled local oscillators ensure better stability and varactor harmonic generators are being used as up-convertors in the r.f. stages.

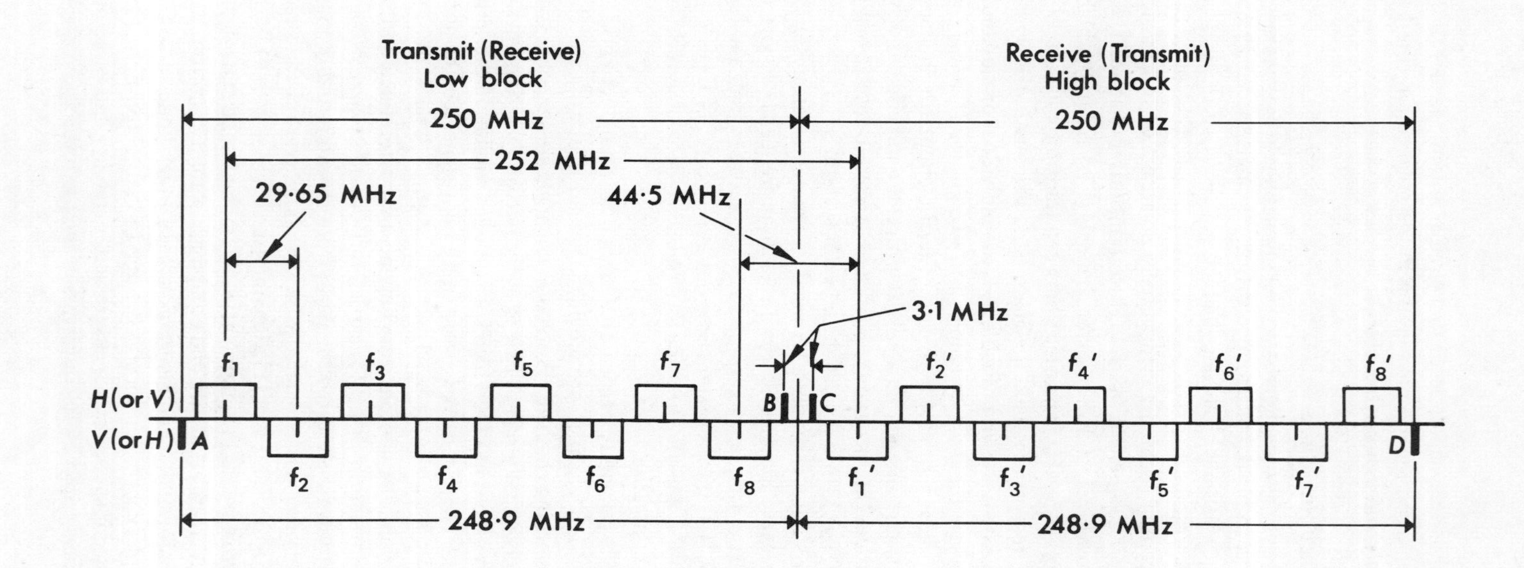

Figure 7.2 Channel assignments for common-carrier frequency band 5·925–6·425 GHz

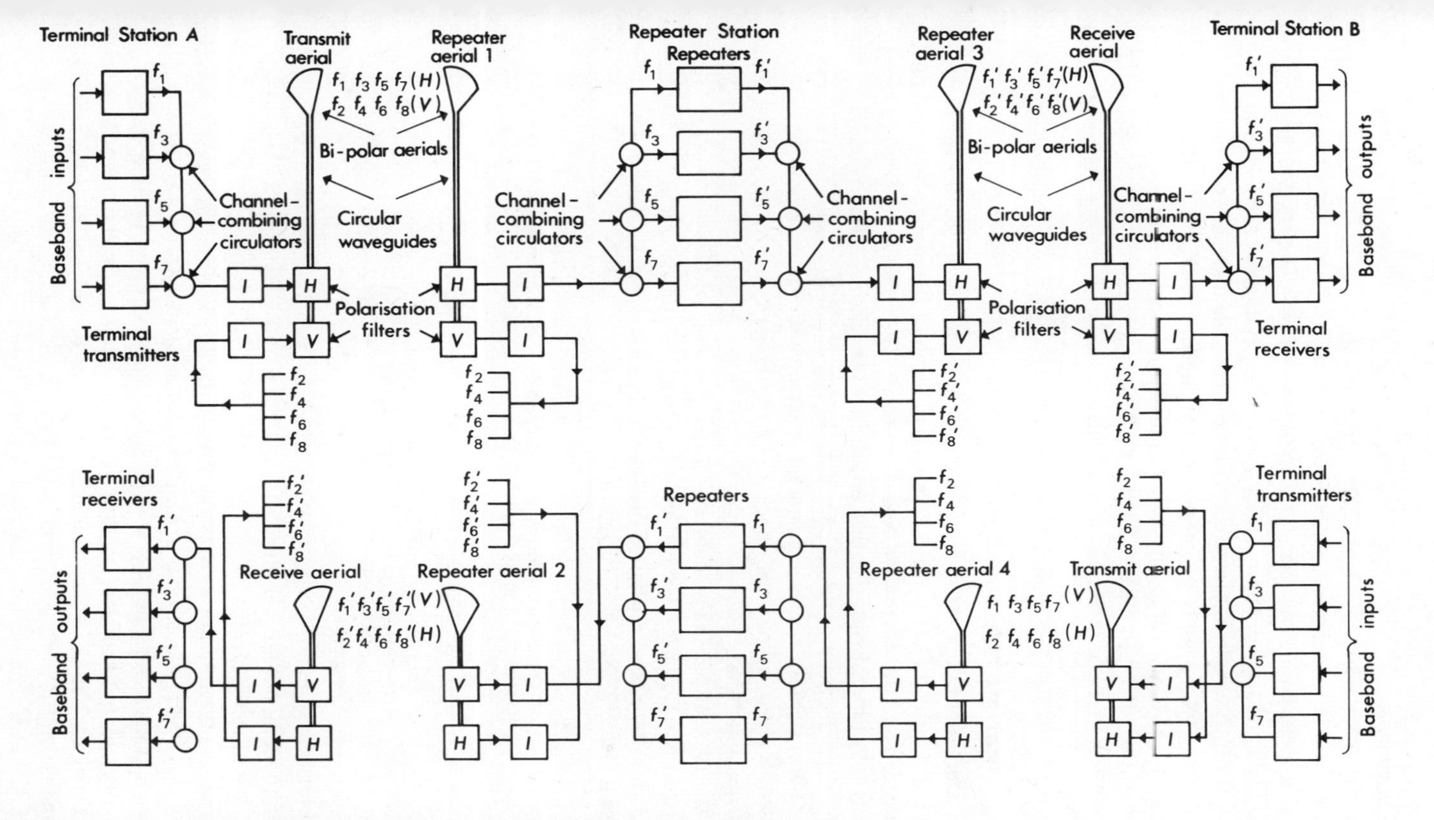

Figure 7.3 Multi-broadband channel microwave system (from ref. 2)

Frequency plans

The frequency plan (e.g. Figure 7.2) determines the number of channels, channel bandwidth, separations etc. and depends on many factors including CCIR specifications on interference, types of service, frequency deviation, fade margin and quality of service. Methods to combat channel interference including the use of *V/H* polarisations and frequency frogging between high and low blocks between offset repeaters are commonly employed. The *CCIR* [1] (Recommendations 283, 382, 387 for example) have published several frequency plans for different bands which take into account all the above mentioned factors and it is upon these that microwave radio equipment manufacturers base their equipment design.

7.3 Microwave system quality

The function of the microwave system is to provide a quantity of bandwidth of a certain quality. The quantity is defined in terms of the number of telephony or video channels the microwave bearer can carry. The quality is defined in terms of the signal-to-noise ratio within one of these telephone or video channels. The latter is the measure of the noise and distortion added to a signal in transmission over the system. In a speech channel, for example, the actual amount of noise present is not of prime importance; the quality is governed by the relative levels of signal and noise.

The noise added to a signal transmitted over a microwave link can be classified into two types, (i) thermal, and (ii) intermodulation noise, both of which are related to the characteristics of the system. For a single radio link the signal to noise is given by

$$S/N = P_T + (G_T + G_R) - (L_T + L_R) - A - (N_T + N_I + N_F) + I + (P + W)$$

P_T	$(G_T + G_R)$	$(L_T + L_R)$	A
Transmitter power (c.w. for f.m. system)	Antenna gain	Equipment losses	Path losses including fading

$(N_T + N_I + N_F)$	I	$(P + W)$
Thermal and intermodulation noise + noise factor of receiver	Modulation improvement factor	Pre-emphasis and weighting advantages

This relation is represented graphically in Figure 7.4 (for a normal link in which intermodulation noise is negligible) where the fading variations of the propagation loss are also shown.

The two types of noise affect the channel quality in different ways. Thermal noise may be broken down into a component generated at the antenna which is dependent upon the input signal level and a component developed in the circuitry, often called 'idle' 'intrinsic' or thermal noise, which is not affected by the signal level. Both of these are independent of system loading. Intermodulation noise on the other hand is affected by system loading, increasing as the traffic increases, but is not in general dependent on the carrier level.

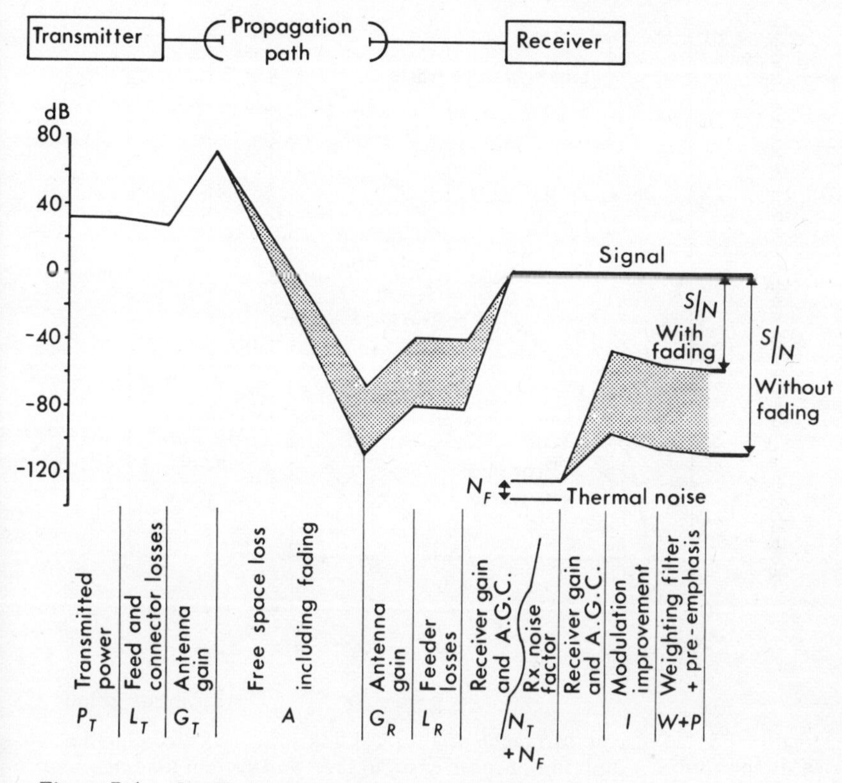

Figure 7.4 Level diagram for a typical Radio-relay link

The various effects of the noise components on the link S/N are illustrated in Figure 7.5, in which the S/N is plotted against the signal input level. The effect of receiver front-end thermal noise is shown by the full line (A) and it is evident from this that in normal operation (input levels –30 to –40 dBm), when the link is undergoing only a shallow fade, and even at the threshold, that this is the controlling noise component. At higher input levels, thermal noise sets a limit to the channel S/N when the system is unloaded (or lightly loaded) and intermodulation noise (B) is the limiting factor for busy-hour conditions. The remaining components of the systems equations are detailed below.

(a) *Path attenuation (A)* For a properly designed link (as described in Chapter 5) the attenuation is merely the free space attenuation,

$$A = 20 \log_{10} \left(\frac{4\pi d}{\lambda}\right) \text{ dB}$$

where d is taken as the distance between the transmitter and receiver, and λ is the wavelength at the band reference frequency (i.e. 2, 4, 6 or 11 GHz). This loss

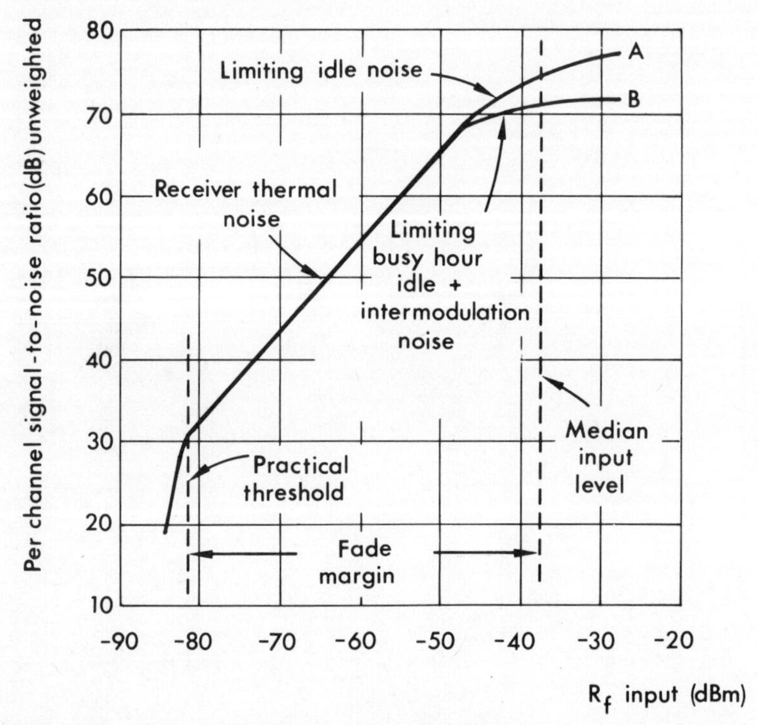

Figure 7.5 Noise performance of a typical link showing worst channel noise plotted as a function of receiver input level and system loading

remains essentially constant over the transmitted bandwidth and may be obtained from the nomograph given in Figure 7.6.

The three most important quantities to be obtained for the link are: practical threshold level, the normal operating level, and the busy-hour operating level. The first two of these are related by the fade margin (which may be calculated as given in Chapter 5) and the third may be obtained by knowledge of the intermodulation noise/thermal noise ratio. Thus essentially the systems calculation need only employ the normal, free-space attenuation loss and derive the other quantities as explained.

(b) *Feeder losses (L)* These arise due to the attenuation in the antenna feeders and insertion losses of associated feeding components. The line or waveguide loss is calculated from the length of the feeder and the characteristic attenuation of the selected feeder, as given by manufacturers data at the chosen frequency.

Microwave filters, isolators and diplexers add loss which is usually counted as part of the feeder loss. Polarisation filters for connecting two waveguide feeders to the same antenna are also considered part of the feeder loss. Typically component losses range from 0·1 to 1 dB and as a rule the total feeder loss should not exceed 3 dB.

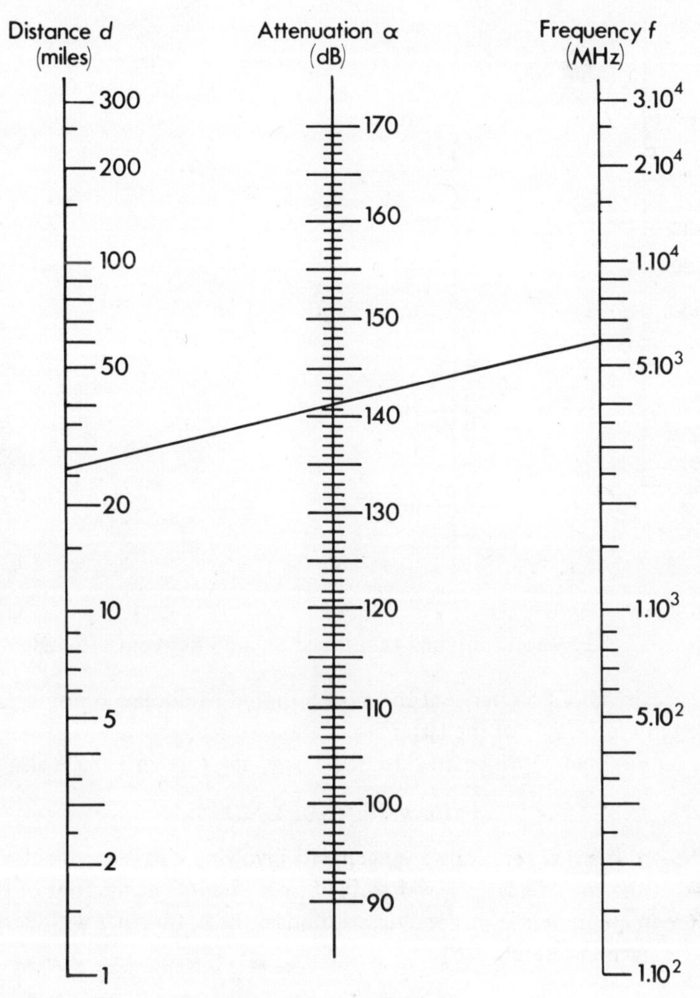

Figure 7.6 Free-space attenuation nomograph

(c) *Antenna gains (G)* The types of antenna commonly used have gains of about 40 dB and are either horn-fed parabolic dishes or horn-reflectors.

Parabolic dishes when used in the bands above 2 GHz are rarely larger than 3 m in diameter. The gain available is greater as the antenna size is increased, because of the narrower beam and increased directivity. The gain of microwave antennas is given by

$$G = \frac{\pi^2 kD^2}{\lambda^2}$$

where D is the diameter of the reflector or aperture, and k is an efficiency

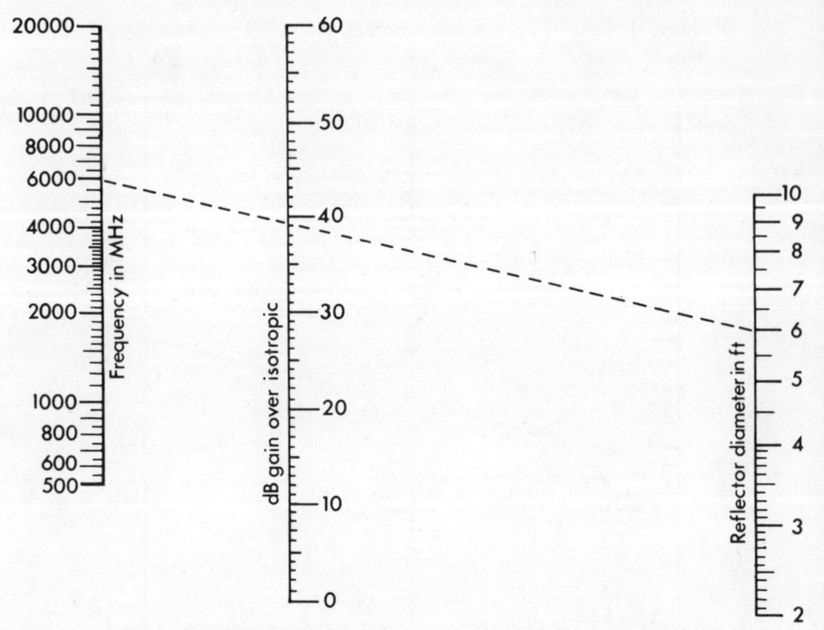

Figure 7.7 Parabolic antenna gain based on a 55 per cent efficiency

factor, usually about 55 per cent for most horn-fed parabolas. A nomogram is included in Figure 7.7 for the latter application.

The approximate 3 dB beamwidth for a large aperture antenna is also given as

$$\text{beam width} = 70^\circ \times \lambda/D$$

In North America periscope arrangements involving a passive reflector situated at the top of a large tower and fed by a parabola at the foot of the tower are in common use. A good description of these, together with design details, is given in reference [6].

(d) *Thermal noise and receiver noise factor* $(N_T + N_F)$ Assuming that the antennas are matched, the front-end thermal noise is given by

$$p = kT_a B \text{ watts}$$

where k is Boltzmann's constant ($1{\cdot}38 \times 10^{-23}$ Joules/K), B is the bandwidth of the receiver in Hz, and T_a is the background temperature in degrees K seen by the antenna.

In microwave radio relay links the antennas are pointed only a few degrees above the horizon and see a background temperature of approximately 300 K. Whence a noise power of -174 dBm/Hz is seen at the receiver input,

$$N_T = -174 + 10 \log_{10} B \text{ dBm}$$

The total receiver noise is the front-end noise plus the noise generated within the receiver, which is given by the receiver noise factor N_F expressed in dB often quoted by the equipment manufacturers (see definition in section 4.4). It is important to note that the noise factor only equals the ratio of output to input S/N's if the input is terminated at 290 K. This is usually taken to be approximately true for radio-relay links. Hence a typical receiver with a bandwidth of 10 MHz has a thermal noise contribution of –104 dBm from the front-end, and the noise generated within the receiver adds to this; with a noise factor of 10 dB, total thermal noise at the receiver output will be –94 dBm.

In the foregoing example an r.f. signal received at –94 dBm would produce a receiver output just equal to the thermal noise. Such a signal would thus not be detectable and this point is known as the noise threshold or a.m. threshold ($-174 + 10 \log_{10} B + N_F$ dBm).

(e) *F.m. improvement factor (I)* Under certain conditions, noise superimposed on an f.m. carrier produces much less effect on the output signal than the same noise on an a.m. carrier. This is because of the characteristics of an f.m. receiver whereby signals with greater peak levels dominate the receiver and suppress lower level signals. If the modulation index (the ratio of f.m. deviation to the applied frequency) of the transmitter is sufficiently large, then the noise can only introduce small phase changes in the signal.

An increase in the modulation index increases the voltage of the demodulated received signal, making the S/N better. Doubling the modulation index (by doubling the deviation) would double the receiver output signal voltage and thus the S/N would increase by 6 dB if the receiver bandwidth were kept constant. However, doubling the deviation would double the bandwidth which the receiver had to accept and thus the noise would be doubled (increased by 3 dB). The net effect is that the S/N is increased by only 3 dB and effectively S/N is directly proportional to deviation.

This characteristic of the receiver is only valid if the wanted signal is dominant. At the noise threshold level, the r.m.s. signal is just equal to the r.m.s. noise power (see Figure 7.8c). However, the peak volume achieved by the noise impulses is approximately 13 dB higher as contrasted with the 3 dB difference between peak-and-r.m.s. values of the r.f. carrier (see Figure 7.8a and b). This is significant due to the feature of f.m. radio that the larger level 'captures' the receiver and suppresses lower ones. Accordingly at the noise threshold (Figure 7.9c) noise is dominant and performances inferior to an a.m. system.

When the level is increased by about 10 dB so that the r.f. carrier peaks equal the noise peaks (Figure 7.8d), the so called *f.m. improvement threshold* is reached (this is the practical threshold). Above this level, the r.f. carrier peaks dominate and S/N increases dB for dB with the carrier level and to the deviation used (see Figure 7.9).

This f.m. improvement is augmented by the use of limiters in the receiver such that a signal well above the noise is limited to a maximum value, and over a wide range of levels the receiver effectively sees only small changes. Typically a

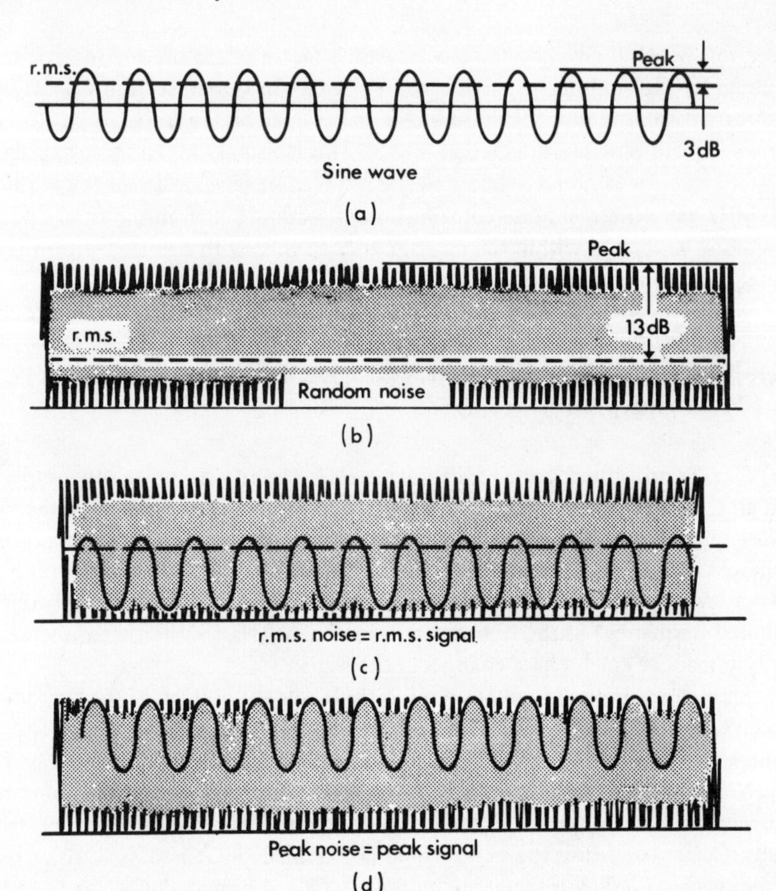

Figure 7.8 Signals in noise (a) r.m.s. power of sine wave signal 3 dB less than peak power, (b) peaks of random noise are 13 dB above r.m.s. power (c) a.m. threshold at which r.m.s. signal power equals r.m.s. noise power (d) f.m. threshold occurs when signal peaks equal or exceed noise peaks, about 10 dB above noise threshold

S/N of 30 dB is obtained at the threshold and the system is designed such that there is only a very small chance of the signal dropping below this in the worst fading condition.

An analysis of the effects of noise on an f.m. system operating in the linear mode is presented in Appendix C, from which it follows that for multichannel telephony the S/N in the top channel (the worst case) is given as

$$\frac{S}{N} = \frac{C}{N} + I$$

where $$I = 20 \log\left(\frac{\Delta f_{\text{rms}}}{f_m}\right) + 10 \log\left(\frac{B_{\text{rf}}}{4}\right)$$

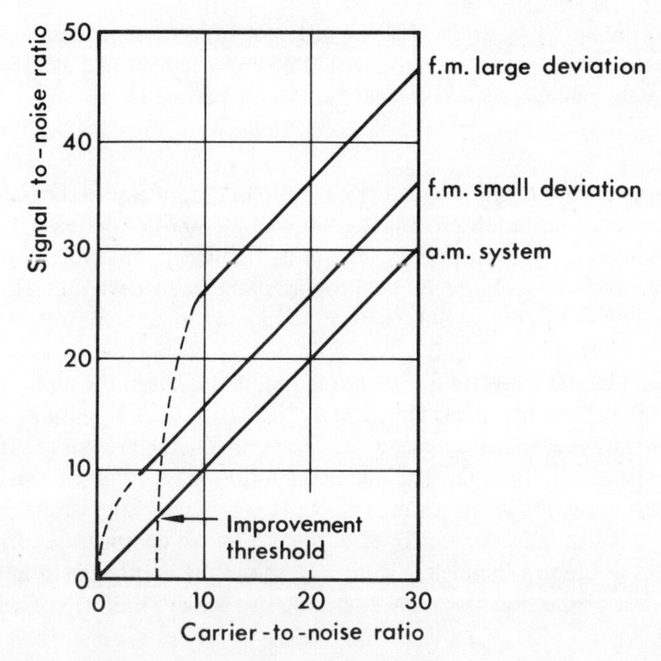

Figure 7.9 Relative performance of a.m. and f.m. systems

where Δf_{rms} is the r.m.s. test tone frequency deviation, which is the frequency deviation produced when a single telephone channel is modulated by a 1 mW tone at the reference point (i.e. OdBmO, see section 2.3), f_m is the top channel modulating frequency and B_{rf} is the bandwidth given by Carson's rule as

$$B_{rf} = 2(F + f_m)\,\text{KHz}$$

where F is the multichannel peak frequency deviation; see (g). It should be noted that the latter bandwidth is often taken as a general rule, but is actually an approximation, albeit a very good one for the majority of practical cases, and the actual bandwidth required may to some extent be a function of the modulating waveform and the quality of transmission desired (i.e. how many f.m. sidebands can be sacrificed).

(f) *Weighting* (*W*) *and pre-emphasis* (*P*)

Weighting

Details of the CCIR recommended psophometric weighting network have already been given in Chapter 1. The improvement factor within a 4 kHz channel is 3·6 dB. Details of the weighting network associated with television transmission are to be found in CCIR documents and the corresponding improvements (for 625 lines/5 MHz channels) are 8·5 dB for flat noise and 16·3 dB for triangular noise, the distribution associated with f.m. N.B. The weighting networks used for 625 line colour T.V. in the U.K. are different from those recommended by the CCIR. The U.K. networks give a weighting improvement of 12·3 dB in the luminance channel and approximately 2 dB in the chrominance.

Pre-emphasis

An f.m. system suppresses idle or thermal noise in proportion to the modulation index (frequency deviation/modulating frequency). Since deviation is a function of the modulating signal amplitude and not its frequency, both high and low frequency baseband components can produce comparable deviation. However, the lower frequency components (higher modulation index) will suppress background noise more. This is why high-frequency channels are noisier than low-frequency ones when transmitted over f.m. radio, unless the modulation index of the higher channels is increased by increasing their amplitude. When this is done, it is called *pre-emphasis* and serves to equalise the noise difference across the baseband. In order that channels are restored to their correct level, a *de-emphasis* network, which has complementary characteristics to the pre-emphasis network, must be employed at the receiver to compensate for the higher level of the top channels. Figure 7.10 shows the improvement in high-channel noise at the expense of lower channels to achieve uniformity across the baseband.

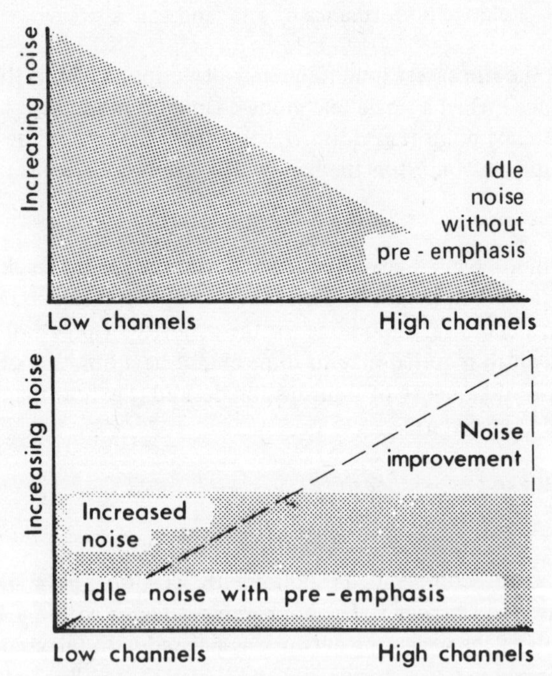

Figure 7.10 Ideal comparison of baseband noise distribution showing how pre-emphasis improves high-channel noise at the expense of lower channels to achieve uniform performance

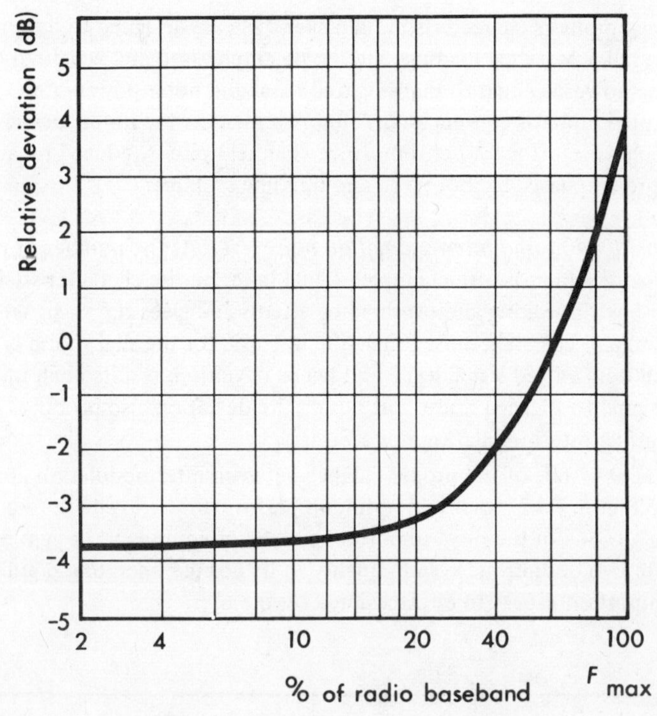

Figure 7.11 Pre-emphasis characteristics recommended by CCIR for multichannel telephony

Figure 7.11 shows the pre-emphasis characteristic recommended by CCIR [1] for multichannel telephony (Recommendation 275). As far as systems calculations are concerned, this means ±4 dB improvement either side of the mean, so that the top telephone channel is increased by 4 dB.

Pre-emphasis is also used for television on f.m. microwave systems, although not for the purpose of noise equalisation. For television transmission the baseband shaping is referred to as pre-distortion rather than pre-emphasis [7].

T.V. pre-emphasis

Unlike an f.d.m. block of telephone channels, a T.V. signal does not have a 'flat' baseband spectrum; most of the energy falls in the low-frequency region, particularly around frame and line frequencies. In order to make the T.V. waveform compatible with telephony therefore, so that equipment of similar design, e.g. frequency modulation modems, may be employed for either signal, it is current practice to include a pre-emphasis network before modulation and a corresponding de-emphasis network after demodulation of the T.V. signal. This has the additional advantage of making the short-term r.f. spectrum more symmetrical about the carrier frequency and thus reducing the problems of automatic frequency control at the transmitter.

The pre-emphasis characteristics in current use are designed to have no effect on the overall T.V. signal loading, and the de-emphasised and weighted baseband noise power is equal to the weighted baseband noise power–i.e. standard pre-emphasis confers no S/N improvement in the luminance channel (or for monochrome). However chrominance channel noise is reduced by about 3 dB by pre-emphasis, for both 525 and 625 line systems.

(g) *Channel loading and intermodulation noise* (N_I) As the number of multiplexed channels increases the complex signal takes on the characteristics of white noise with peaking and overloading effects and gives rise to signal-to-intermodulation and crosstalk noise. Since the intrinsic (or thermal) noise is fixed, raising the transmitted signal level, and hence deviation, results in an improvement of signal to thermal noise. But raising the deviation also produces an increase in the intermodulation noise.

The maintenance of the proper balance between intermodulation and thermal noise (see Figure 7.12) to obtain optimum performance with these two conflicting factors, is one of the most important aspects of equipment design. Since the sum of the two defines the system quality in its normal operating condition, it is also an important aspect to be considered by the user.

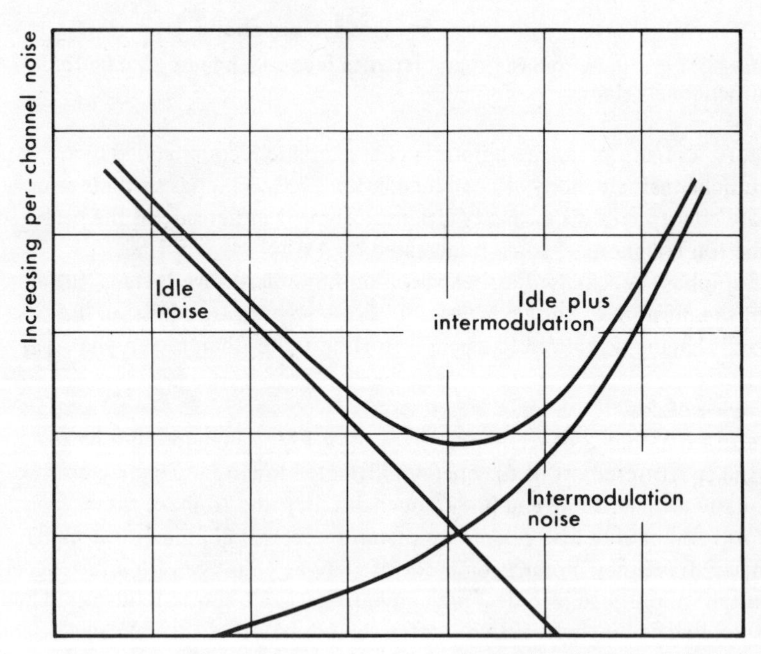

Figure 7.12 Noise performance of an f.m. system, showing optimum deviation level

Channel loading
As explained in section 2.3 it is useful to refer powers in the transmission system to a reference point (dBm0). In the f.m. system it is necessary to develop a relationship between the r.m.s. frequency deviation of the signal and the power of that signal at the reference point.

In order to do this use can be made of the relationship between the multiplex signal and the peak frequency deviation which the signal produces. Chapter 4 discusses and shows how to determine the equivalent r.m.s. load, L_N, which a multiplexed telephone system must be designed to carry. This is given as a power referred to the reference point, and thus the corresponding frequency deviation must be normalised accordingly in order to equate them, i.e. for a linear frequency demodulator,

$$20 \log \left(\frac{F_{\mathrm{rms}}}{\Delta f_{\mathrm{rms}}} \right) = L_N$$

where F_{rms} is the r.m.s. deviation of the multichannel load and Δf_{rms} is the r.m.s. test tone frequency deviation defined as the frequency deviation produced when a single telephone channel is modulated by a 1 mW tone at the reference point.

N.B. L_N is often referred to as the multichannel r.m.s. load in radio work and is the ratio of the multichannel r.m.s. voltage to the r.m.s. channel test tone voltage.

The transmission bandwidth involves the peak multichannel signal deviation which is given as

$$F = cF_{\mathrm{rms}}$$

where c is the peak voltage to r.m.s. voltage of the multichannel load which is not exceeded for a percentage of the time. This varies with the number of channels (see P.92) and for radio relay work is usually taken as 13 dB.

The peak frequency deviation of the multichannel signal may also be given as

$$F = cl\Delta f_{\mathrm{rms}}$$

where $$L_N = 20 \log l$$

Intermodulation
The effects of transmission non-linearities in amplitude and group delay in producing intermodulation amongst the channels of an f.d.m. signal have already been discussed in section 4.2.

Intermodulation due to non-linear filters and amplifiers is in general calculable [7] although outside the scope of this text. Group delay of channel filters is particularly troublesome and this is usually quoted with a linear and parabolic component which must not exceed a few nanoseconds in a 30 MHz channel to avoid signal distortion. In general equalisation of all equipment group delay degradation is performed in a single device incorporated in the receiver.

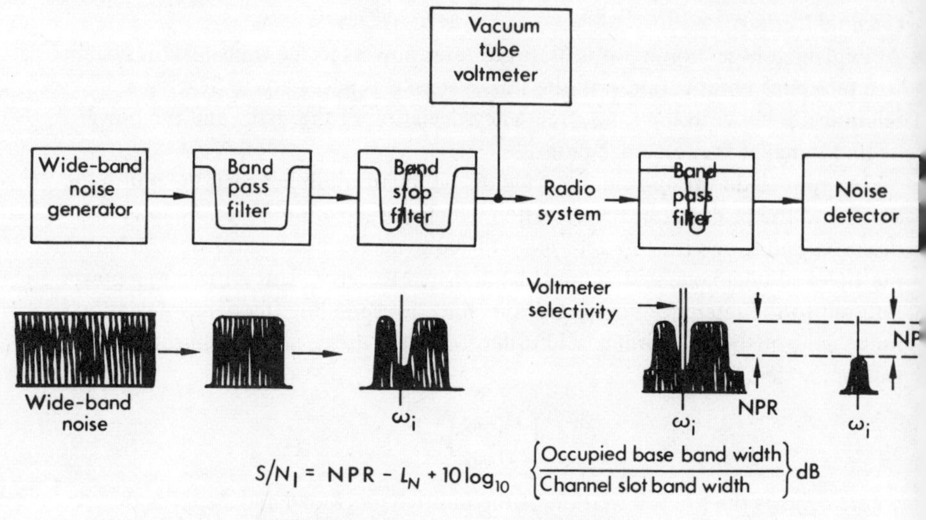

Figure 7.13 Intermodulation noise measurement

Also of particular nuisance value are echos, particularly at the antennas, which constitute signal distortion [8]; they may be alleviated by liberal use of isolators. Intermodulation noise comprises the sum of many independent noise sources, the two most important of which have been mentioned.

It becomes fairly important for the system user to be able to evaluate the signal-to-intermodulation noise under busy conditions. The CCIR have recommended (No. 399) a procedure for doing this which makes use of the equivalent white noise loading [8] for a multichannel signal (explained in Chapter 4) and which is shown diagrammatically in Figure 7.13.

Also available nowadays are radio systems link analysers, which measure amplitude, frequency and group delay characteristics at i.f. for both multichannel telephony and television reference signals.

Overall radio system noise objectives

CCIR have recommended (No. 395) a noise power not exceeding 10 000 pW0p for a 2500 km circuit for telephony. This corresponds to a signal-to-noise of 50 dB referred to the OdBmO point. 1 pW/km is allowed for multiplex equipment and thus the planning objective is 3 pW/km, as for a cable system (see Chapter 3).

Multihop performance

The discussion so far has concerned only the performance of a single hop. In a single-hop system, the limiting factor in system performance is generally the noise contribution of the multiplex channels operating over the radio rather than the radio noise. But in a long microwave system noise is contributed by each individual hop. With every hop operating normally, i.e. all paths unfaded, the thermal and intrinsic noise will increase as the number of hops increases. Their total noise power can be obtained by power addition of the single hop noise. Thus for a six-hop system, thermal noise will increase by $10 \log 6 = 7{\cdot}8$ dB.

Intermodulation products, on the other hand, while they are essentially random for a single hop, do have some coherency. If the intermodulation products were coherent, adding in phase on each hop, then the result would be voltage addition and result for a six-hop system in an increase of $10 \log 6^2 = 15{\cdot}6$ dB.

In fact, only odd order intermodulation products are coherent and the actual effect is somewhat better. A typical figure based on empirical results, is obtained by using $15 \log N$ for intermodulation noise, where N is the number of hops.

7.4 Radio systems planning example

In this simple example the mathematical parts of the chapter are exemplified in a systems planning example. Obviously more considerations exist than are brought out here, as systems planning is not always a straightforward procedure and all of its steps cannot be covered by mathematical formulae. Many trade-offs between cost, maintenance difficulties, time scales etc. rely to a great deal on the experience of the designer, but in general he must provide a system which meets certain transmission and reliability objectives. Early design calculations are merely aimed at comparing the performance of possible arrangements, one of which is given here, and many iterations may be made before figures are agreed upon and the complicated process of path planning begins, in which repeater sightings are made after extensive path surveys.

System objectives

For the purposes of this illustration it is proposed to install a new 6 GHz trunk facility to replace some outdated 2 GHz equipment. Some preliminary data is as follows:

(a) The length of the system is 320 kilometres with 8 hops.

(b) The system should be planned for 99 per cent reliability and to have a busy hour noise level under these conditions not exceeding that specified by the CCIR. It should also be capable of satisfactory operation with a nominal 40 dB fade on any hop.

(c) The system should be capable of transmitting 1800 telephone channels and television.

(d) All existing plant should be used where appropriate.

Design procedure

The first step is to decide upon a frequency plan and to some extent the designer is constrained to CCIR recommendations, and compatibility with existing systems. For these reasons the design will conform with the widely existing commercial equipment which is based on the CCIR recommendation No. 383

for 1800 channel systems (Figure 7.2). The channel frequency spacing used here is 29·65 MHz.

1800 channels in f.d.m. multiplex occupy the baseband region 312 kHz-8·20 MHz (see chapter 4).

The r.f. bandwidth required to transmit the block has already been given from the frequency plan but is also given from Carson's rule as

$$B_{rf} = 2(cl\Delta f_{rms} + f_m)$$

where c, the peak to mean ratio, 13 dB (or 4·47), and f_m, the top modulating frequency, 8·204 MHz, have been determined. Hence

$$\begin{aligned} 20 \log_{10} l &= L_N \\ &= -15 + 10 \log_{10} (1800) \text{ dBm0} \\ &= 17{\cdot}6 \text{ dBm0} \end{aligned}$$

Hence $$l = 7{\cdot}586$$

Thus the r.m.s. test tone deviation is calculated as

$$\Delta f_{rms} = \frac{29{\cdot}65/2 - 8{\cdot}204}{4{\cdot}47 \times 7{\cdot}586}$$

whence $$\Delta f_{rms} = 195 \text{ kHz}$$

A typical hop power budget is now estimated.

	Losses	*Gains*
Transmitter power		P_T dBm
Antenna gains		80 dB
Feeder losses	2 dB	
Flexible cable losses	0·5 dB	
Filter/diplexer losses	0·5 dB	
Free space loss	142 db	
	145 dB	P_T + 80 dBm

The unfaded carrier level is (P_T – 65) dBm.

The next design step is to allocate the systems noise objective which is according to CCIR allocations

3 x 320 = 960 pW0p transmission path
1 x 320 = 320 pW0p multiplex equipment

An arbitrary division is made in Figure 7.14 which assumes that the noise contributors add on a power basis. This would normally be done with performance estimates of existing equipment to hand and the remainder divided equally between thermal and intermodulation noise, both of which limit the system performance. If any of the individual block noise performances cannot be attained then the division would be rearranged accordingly.

The idle noise objective of 450 pW0p (63·5 dB referred as a S/N to the

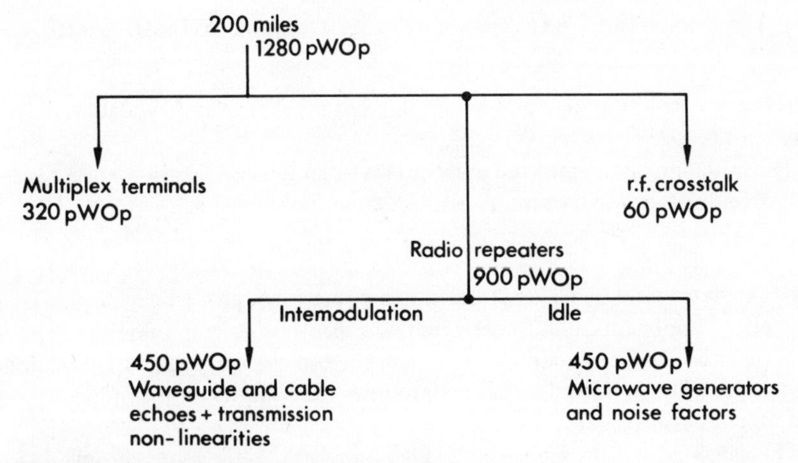

Figure 7.14 System noise allocations

0 dBm0 point) may be translated to a single hop figure via

$$63{\cdot}5 + 10 \log 8 = 72{\cdot}5 \text{ dB}$$

The system reliability of 99 per cent to account for a cumulative fade would require each hop to be 99·99 per cent reliable and, assuming a properly engineered path, a Rayleigh fade of 28 dB (see Figure 5.16) is allowed for in the path attenuation. Thus the hop transmission equation is

$$S/N = C/N + I + P + W$$

$$\text{Faded carrier} = (P_T - 65 - 28) = (P_T - 93) \text{ dBm}$$

$$\text{Receiver noise} = -174 + 10 \log_{10} (2{\cdot}965 \times 10^7) + N_F$$

$$= (-99 + N_F) \text{ dBm}$$

$$\text{Modulation improvement } I = 20 \log_{10} \left(\frac{\Delta f_{\text{rms}}}{f_{\text{m}}}\right) + 10 \log_{10} \left(\frac{B_{\text{rf}}}{4}\right)$$

$$= 20 \log_{10} \left(\frac{0{\cdot}195}{8{\cdot}204}\right) + 10 \log_{10} \left(\frac{2{\cdot}965 \times 10^4}{4}\right) = 6{\cdot}3 \text{ dB}$$

Hence
$$72{\cdot}5 \leqslant (P_T - 93) + (99 - N_F) + 6{\cdot}3 + 4 + 3{\cdot}6$$

$$52{\cdot}6 \leqslant P_T - N_F$$

Practical noise factors for microwave receivers are between 8-10 dB and in this case implies a transmitter power of between 60·6 and 62·6 dBm.

Taking the 10 dB figure, the device required in the transmitter would be a travelling wave tube amplifier with output level 63 dBm.

The next step is to test for the threshold (or breaking point) on a faded hop. This is done by comparing the carrier level during a 40 dB fade with the total thermal noise referred to the receiver input in a channel bandwidth of 29·65 MHz.

$$\text{Threshold} = (63 - 65 - 40) - (-174 + 10 \log_{10} (2{\cdot}965 \times 10^4) + 10)$$
$$= -42 + 89 = 47 \text{ dB}$$

Thus breaking will not be a problem in this system as serious degradation only results if the C/N drops to 10-15 dB.

The initial breakdown of the noise results in an intermodulation noise of 450 pW0p for the whole route. A per hop value is obtained as

$$63{\cdot}5 + 15 \log_{10} 8 = 77 \text{ dB}$$

The next step would be to break this down into echoes, group delay degradations etc., components, and to perform noise spectrum calculations [7] to generate curves of intermodulation noise in the top channel versus transmission degradations. An iterative process then follows to set limits on the individual component performances.

The weighted signal-to-noise on a faded hop for 625-line monochrome television transmission is calculated, with reference to CCIR recommendations:

$$S/N \text{ (faded hop)} = C/N + 20 \log \left(\frac{\sqrt{3}\, F_{pp}}{f_m}\right) + 10 \log \left(\frac{B_{rf}}{f_m}\right) - 3 + 12{\cdot}3$$
$$= 41 + 20 \log \left(\frac{8\sqrt{3}}{5{\cdot}5}\right) + 10 \log \left(\frac{30}{5{\cdot}5}\right) + 9{\cdot}3 = 65{\cdot}9 \text{ dB}$$

when this noise is added to the noise from the rest of the system, the sum is within the CCIR objective of 56 dB.

Such systems calculations, especially when iterations are necessary, could be made much easier by the use of digital computing techniques, this being a relatively new innovation which is currently receiving attention [4, 9].

7.5 Troposcatter sections

In certain locations the use of line-of-sight chains is uneconomic or even impossible due to unfavourable terrain in underdeveloped countryside where the building of access roads and the maintenance of remote repeaters poses problems. In these cases several line-of-sight hops in the route may be replaced with an over-the-horizon or troposcatter link. The feasibility of this approach depends on whether the long link can be realised with reasonable expenditure, preferably with antennas not exceeding 10 m diameter, and with transmitter powers of not more than 1 kW. To make good use of available antennas, the region 900 MHz-3 GHz is usually chosen for operation of these links and the receivers often include the more sensitive tunnel-diode amplifier or parametric amplifier. Quadruple diversity is a necessity due to the propagation mechanism (see Chapter 5) and to achieve the required reliability, greater than 99 per cent in the worst propagation month.

Systems design is essentially the same as in section 7.4, except that reliance on empirical data for the path attenuation is necessary, and this is usually contributed by a period of actual path testing before commencement. All systems design should be made in accordance with CCIR [1] Recommendations (Nos. 286, 388 and 296, 297) and reference 10 is an accepted planning manual with most authorities.

Regarding the frequency plan it has to be kept in mind that separate frequencies are required for all transmitters, otherwise mutual interference would arise due to over-reach propagation. The frequency spacing between adjacent channels should be at least 50 MHz in order to effect the desired frequency diversity effect. The transmit/receive frequency spacing has to be at least 100 MHz for protection of receiver input against the local transmitter. Otherwise no more than 1·5 MHz spacing is required between the r.f. channels of the systems if crystal control is provided for all frequencies.

REFERENCES

1. CCIR, *Documents of XIIth Plenary Assembly*, New Delhi 1970, Part IV, Part 1, 'Fixed Services Using Radio-Relay Systems'.
2. D. Jones and P. Edwards, 'The Post Office Network of Radio Relay Stations', *Post Office Electrical Engineers Journal* **57**, October 1964, p. 147 and January 1965, p. 238.
3. 'The Post Office Tower, London, and the United Kingdom Network of Microwave Links', *Post Office Electrical Engineers Journal* **58**, 3, October 1965, p. 149-159.
4. 'All Semiconductored Microwave Radio-Relay Systems', *G.E.C.-A.E.I. Telecommunications*, No. 38, 1970.
5. S. Hathaway *et al.*, 'The TD3 Microwave Radio-Relay System', *Bell System Technical Journal*, **47**, September 1968, pp. 1143-1188.
6. *Selected Papers from the Lenkurt Demodulator*, The Lenkurt Electric Company, 1965, p. 313.
7. *Transmission Systems for Communications*, Bell Telephone Laboratories, 4th Edition, Chapter 21; and H. Carl, *Radio-Relay Systems*, MacDonald, London, 1966, Section 4.3.
8. R. G. Medhurst, 'Echo Distortion in Frequency Modulation', *Electronic and Radio Engineer*, July 1959, pp. 253-259.
9. M. Massaro, 'Performance Analysis of Microwave Communications Systems by Digital Computer', *Microwave Journal*, Volume 13, No. 8, August 1970, pp. 35-41.
10. National Bureau of Standard, *Technical Note No. 101.*

Chapter 8

Satellite communication systems

8.1 Introduction

To date the benefit of satellite communication systems has been largely in their cost independence of length. Thus high capacity international traffic has been the main user. The use of three geostationary satellites (36,000 km equatorial orbits) is sufficient to cover the globe[1] and any two locations on the earth can be linked with only two repeaters. The radio system configuration is not unlike those described in the last chapter except for the extra complexity of the satellite repeater and the extra sensitivity of the receiving earth stations. In fact the disadvantages have largely been connected with the cost of the equipment to ensure sufficient reliability and the cost of the launch, especially when a heavy investment in undersea cables has been made. These economic disadvantages have today been invalidated and have left a single remaining drawback—that of delay in a two-way telephone circuit.

Due to the distances involved between satellites in the geostationary orbit and earth stations, a delay of approximately 270 ms exists for a one-way circuit. In a two-way telephone circuit this represents a pause of nearly 0·5 s between speaking and receiving the answer. It is generally assumed that this is subjectively acceptable but the connection of two links in tandem would not be. However, recent investigations[2] have shown that delay alone is tolerable but accompanied by echos of the sort experienced on satellite links considerably reduces the magnitude of the acceptable delay. It is certainly possible that advances in adaptive echo suppressors could in the future permit the use of two links in tandem and hence allow two-way telephone circuits to operate globally via satellites.

8.2 The INTELSAT system

Due to the possibilities of inter-continental communication, it was obvious that an international body was required to promote and co-ordinate a global system. To fulfil this need INTELSAT [3] was formed in 1964 with responsibility to operate and maintain the global system, financed by the countries who are members of the consortium. INTELSAT has vested the management and operation in COMSAT, a commercial company which also represents the United States in INTELSAT. INTELSAT via COMSAT undertakes the design and production of satellites and arranges for their launching (via NASA), tracking and control plus other support facilities. The participant nations arrange their own earth stations which, however, must conform to INTELSAT requirements for the system.

INTELSAT has planned a global system based on stationary synchronous equatorial orbit satellites, which has been accomplished via the phases INTELSAT I to IV satellites placed over the Atlantic, Pacific and Indian Oceans to give a completely global coverage. At the time of writing there are 50 earth stations in 30 countries, and many additional countries obtain service from them. By the end of 1972, the system will have grown to some 86 earth stations in 59 countries.

Besides the INTELSAT system, the USSR operates a substantially national system via the Molnya series of polar orbiting satellites which are used primarily for television transmission to remote areas. There are also many military and navigational systems in operation today.

F.d.m.a.–f.m. operation

The global system, up to and including the INTELSAT III series of satellites, has standardised on the use of f.m. techniques, partly due to the fact that these techniques are well established in terrestrial systems, and partly because efficient use of the satellite power requires a modulation method which spreads the information over a wide band to achieve a good S/N with minimum satellite carrier power. The disadvantage of f.m. is that it also limits the satellite capacity in a power limited situation.

The frequency band allocated by the CCIR for satellite communication is:

5·925-6·425 GHz–earth-to-satellite
3·700-4·200 GHz–satellite-to-earth

with a bandwidth of 500 MHz, which due to the separation requires a direct frequency conversion of 2·225 GHz by the satellite repeater.

These frequency bands are shared with terrestrial users and thus the risk of mutual interference is high. This has resulted in a limitation being placed on the radiated power from all satellites [4]. Hence the need in the future to look towards higher frequency bands.

The satellites act as relay stations, amplifying and retransmitting the signals they receive, so that all earth stations in sight of the satellite can exchange signals indirectly with one another. Simultaneous use of a satellite by several earth stations is called 'multiple access' and this is achieved via the use of multidestination carriers. In this scheme, telephone channels are frequency modulated with frequency division multiplex basebands (f.m.-f.d.m.) onto pre-assigned carriers and transmitted from an earth station regardless of destination, in a 'broadcast' fashion. Every distant station receives all the carriers, but only selects the appropriate carriers or particular channels from one carrier, addressed to it, for demodulation in the multiplex equipment. Stations are

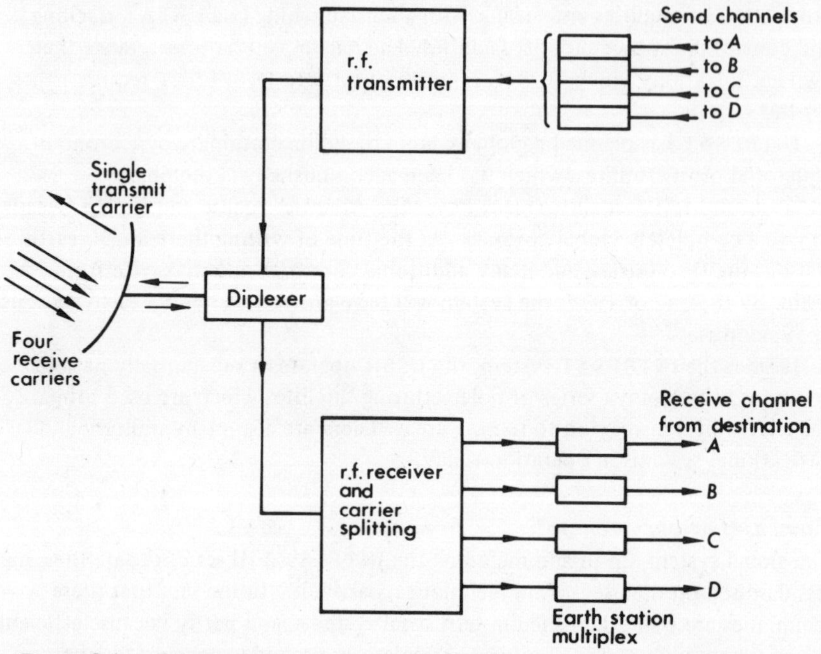

Figure 8.1 Schematic arrangement of transmit and receive carrier for multiple-access working

allocated carriers according to their capacity requirements by INTELSAT, in blocks of 24, 60 or 132 telephone channels or one television video channel [4]. Typically an earth station may have up to three such carriers with specified transmit frequencies fixed by INTELSAT. Such a station could transmit 132 channels destined for 10 countries on one carrier and receive the complementary return paths on 10 separate carriers. Figure 8.1 illustrates the transmission and reception of 4 blocks of channels on this basis. The total capacity of the satellite is about 1200 two-way circuits. Due to the problems of frequency assignment, each earth station must be capable of receiving the full 500 MHz bandwidth as the required carriers may fall anywhere in this band. Thus, although economies may be made on the transmit side, each participatory earth station is stereotyped to a standard configuration (dish size of approximately 27 m).

Satellite transponders

A simplified diagram of the type III satellite transponder arrangement is shown in Figure 8.2. Two separate transponders (225 MHz bandwidth) cover the 500 MHz transmit and receive bands each sharing a common despun horn earth coverage antenna (beamwidth 17°). The 6 GHz signals which are received from the earth station pass through tunnel-diode amplifiers, then after frequency changing to 4 GHz and two stages of travelling wave tube amplification, are radiated from the same horn antenna. The total gain of the transponder is

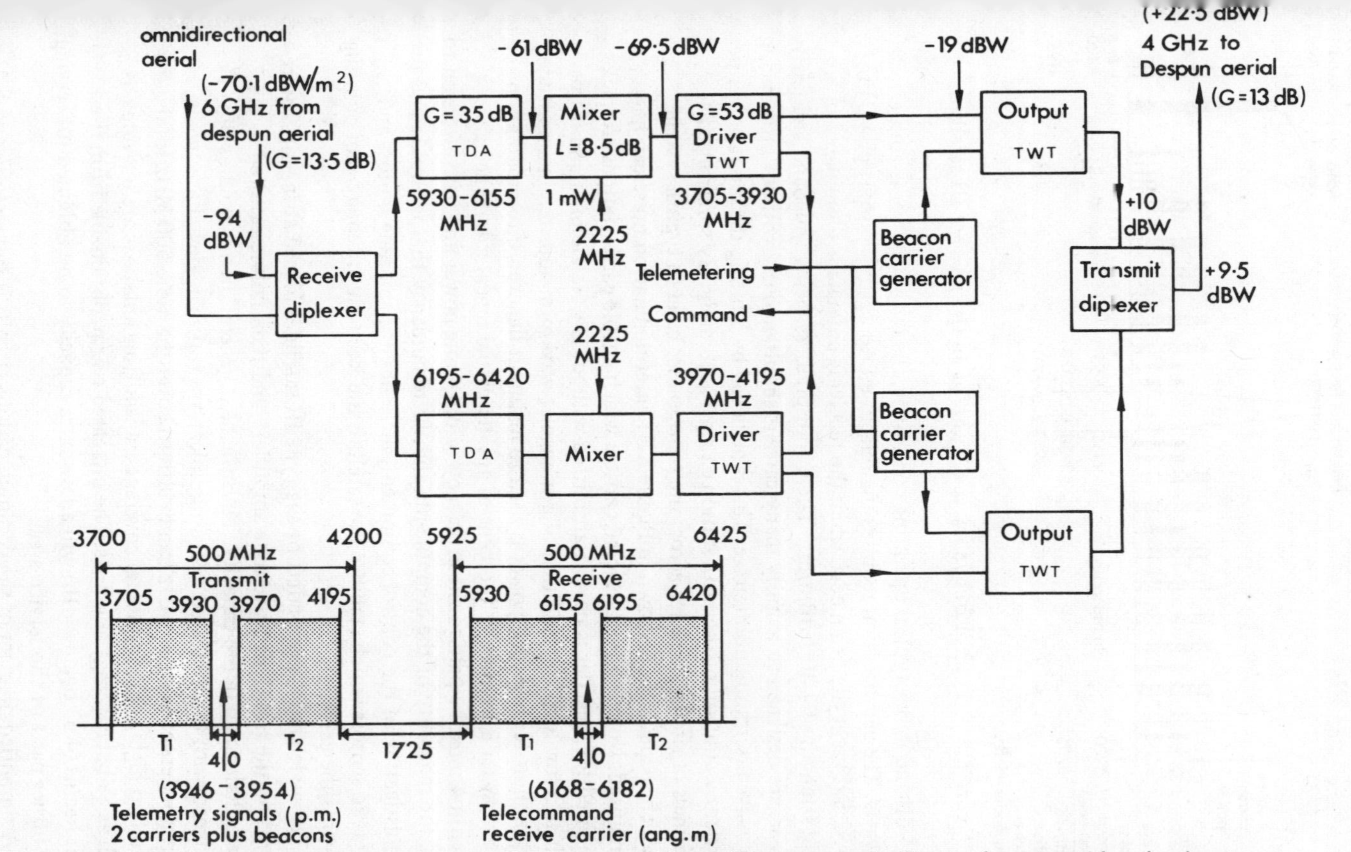

Figure 8.2 Simplified block diagram of INTELSAT III satellite communications sub-system, showing inset transponder frequency plan (N.B. 1 dBW = 30 dBm)

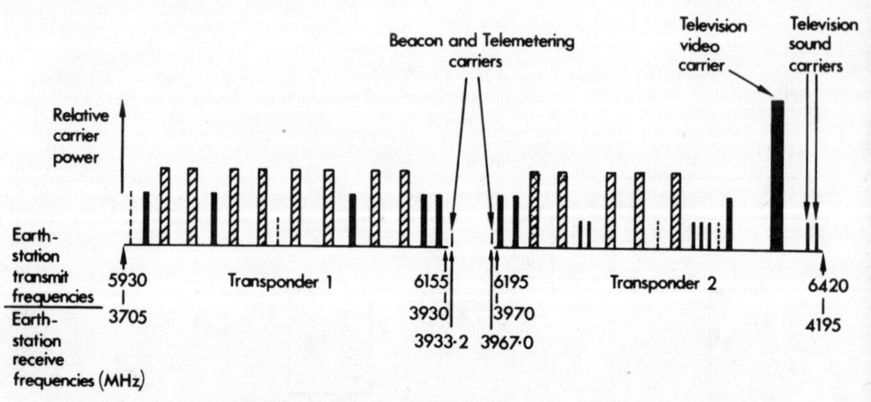

Figure 8.3 Typical frequency spectrum for an INTELSAT III satellite

104 dB and under typical operating conditions, the output power is about 6 watts. The specification requires the output e.i.r.p. (gain x output power) to be 52·5 dBm at the beam edge when the receiving antenna is illuminated by a field strength of –40·1 dBm/m^2. These transponders, being wideband and fairly linear, are designed to amplify many carriers simultaneously without introducing excessive intermodulation noise. This is achieved by limiting the input power per carrier by controlling the earth station e.i.r.p.† accurately. By spreading the channels of similar capacity across the transponder band (a typical system frequency spectrum is shown in Figure 8.3), intermodulation noise can be minimised. Note the two 'beacon' carriers which enable earth stations to track the satellite. A carrier is never transmitted without any traffic on it and if this falls to a low level, the addition of a triangular waveform adjusted in amplitude to the short-term speech power is used to maintain the even intermodulation. The repeater is essentially the same as any terrestrial repeater but special attention must be paid to the efficiency of obtaining power and its conversion to r.f., the reliability of the components and the redundancy that is built in and the sophistication of the control equipment, including the spinning motors and infra-red sensors which stabilise the satellite and keep the antenna beam pointing towards the earth.

The wideband transponders of the type III satellites placed stringent requirements on the travelling wave tube amplifiers, and their failure resulted in the loss of half the satellite capacity. Thus the next generation of satellites type IV, designed with the increased telecommunications traffic in view, have twelve transponders each of 36 MHz bandwidth to cover the same 500 MHz total band. Also two high-gain continental coverage transmitting antennas are provided as well as global coverage antennas. The increased e.i.r.p. obtainable from these spot beams (4·5° beamwidth) will allow extra capacity to be obtained for serving the busier parts of the service area.

A simplified diagram of the communications circuits is shown in Figure 8.4,

†effective isotropic radiated power.

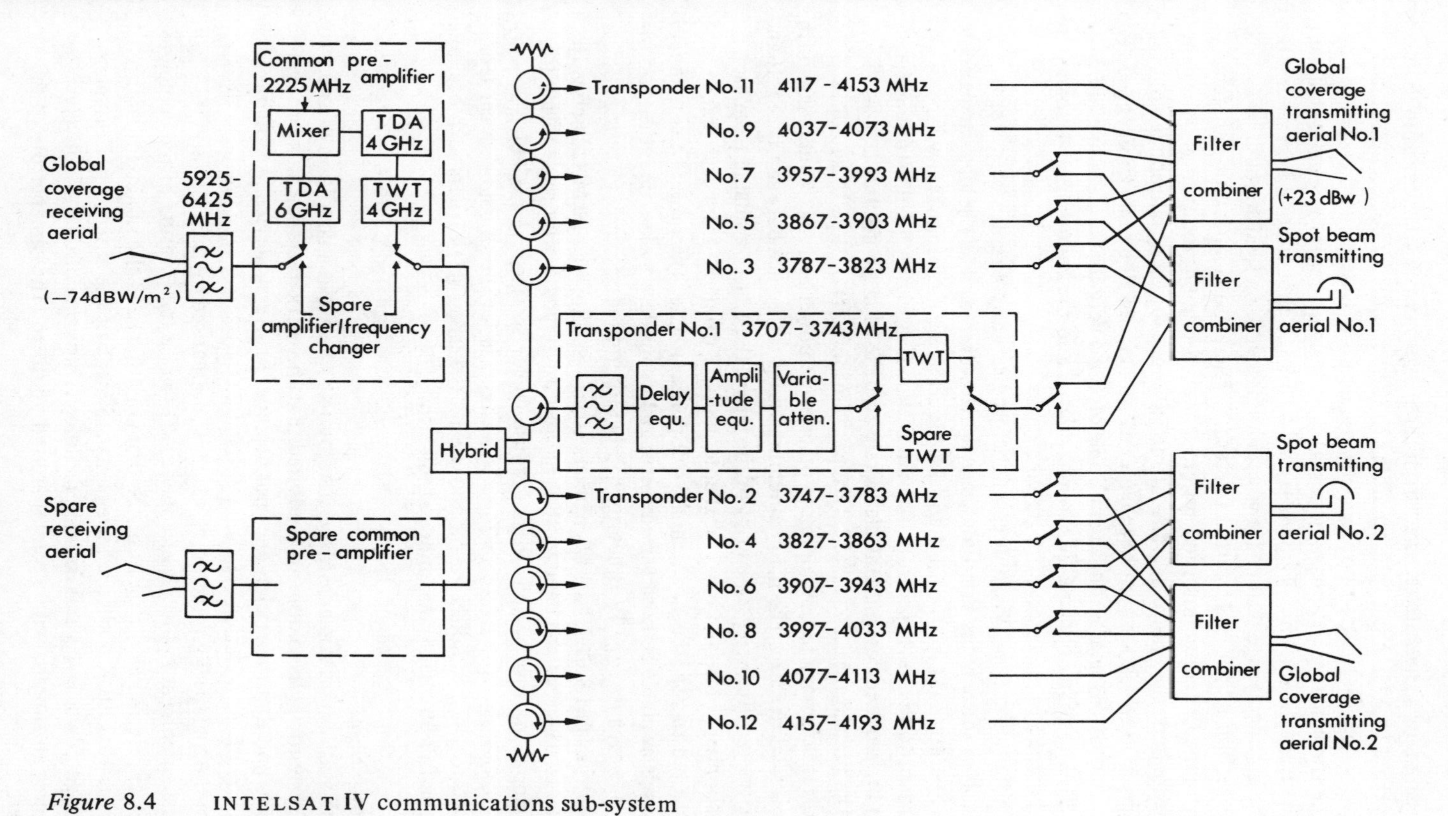

Figure 8.4 INTELSAT IV communications sub-system

showing how signals are received from earth stations at 6 GHz, are amplified, converted to 4 GHz and split according to their frequency between various transponders. In the transponders they are amplified again and then fed to one of the transmitting antennas, for global or spot beam coverage. The overall gain of the transponder (minus antennas) is 134 dB and the output e.i.r.p. is 53 dBm at the earth coverage beam edge when illuminated by a field strength of –43·7 dBm/m^2 for multi-carrier operation. There are facilities for remotely controlling the communication system from the ground; for example, the gain of transponders, the choice of transmitting antennas for each transponder and the direction in which the spot beam antenna points, can all be telecommanded. The new innovations apart from the multiplexer are a very low noise tunnel diode amplifier enabling higher C/N's to be achieved and the inclusion of all solid state equipment and microwave integrated circuits.

Initially the type IV satellites will be used as a direct replacement for the types III's for relaying f.m.-f.d.m. signals. However, as new digital systems come into operation (SPADE for instance, see section 8.5) transponders will be allocated to them. A spectrum diagram is shown in Figure 8.5 illustrating the kind of frequency assignment plan likely to be used in the future.

The limited total e.i.r.p. (52 dBm per transponder) of type III satellites makes it necessary to use minimum practicable power and maximum bandwidth for each carrier size. The main communications difference between types III and IV is the greatly increased e.i.r.p. of the latter; the e.i.r.p. available from each transponder connected to a global antenna is 52 dBm and from each transponder connected to a spotbeam antenna 64 dBm is available. The ratio of the number of channels available is, however, unfortunately considerably less than the ratio of e.i.r.p.'s because type IV is heavily bandwidth limited.

With INTELSAT III a 132 channel carrier was allocated 20 MHz bandwidth; for INTELSAT IV the bandwidth is reduced to 10 and 5 MHz respectively for the global and spot beam carriers. Additional carrier sizes are available for INTELSAT IV working, from 24 to 972 channels for the global beam carriers and 60 to 1872 channels for the spot beam, the larger size carriers each occupying the full 36 MHz transponder bandwidth.

Earth stations

Earth station performance is specified by INTELSAT as a condition of entry to the system[5]. In so doing it is ensured that the CCIR recommendations are upheld and that the station does not degrade the overall service. A suitable measure of earth station efficiency is its ability to detect very low level signals from the high background noise which is summarised by the G/T ratio (G = gain of the antenna at the receive frequency, T = noise temperature of the complete receiver referred to the input of the first amplifier) and the minimum requirement is 40·7 dB/K. Other requirements are made of the antenna performance, radiated power levels and tracking facilities. This value of G/T effectively requires an antenna with a gain of around 58 dB and thus an aperture of the

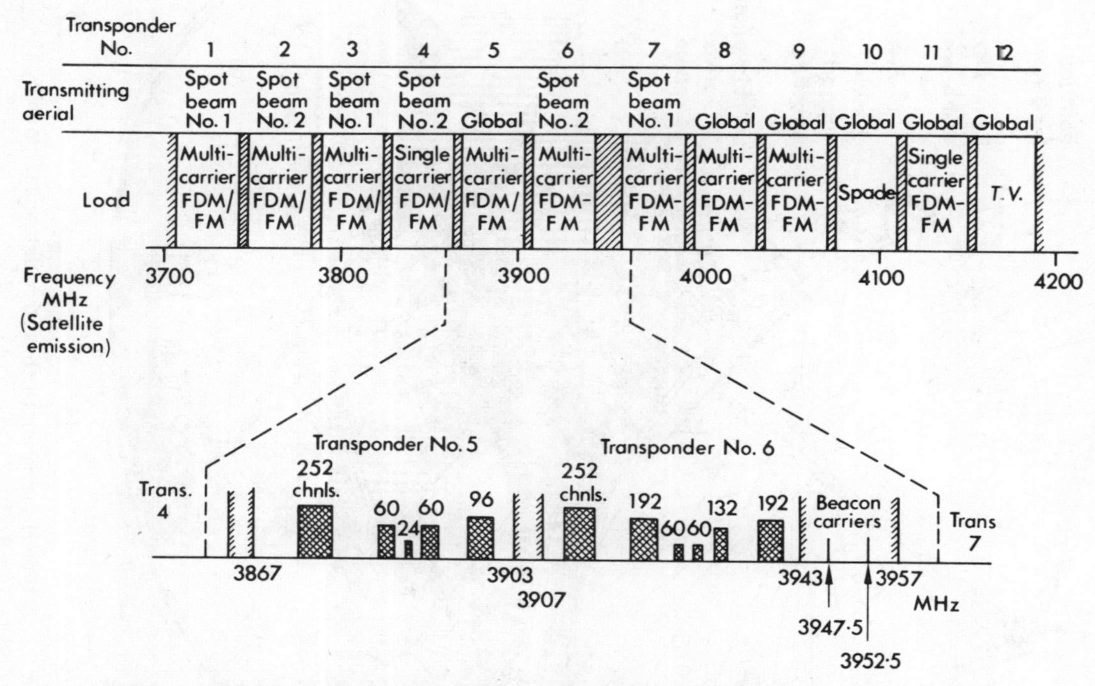

Figure 8.5 A model frequency spectrum for an INTELSAT IV satellite

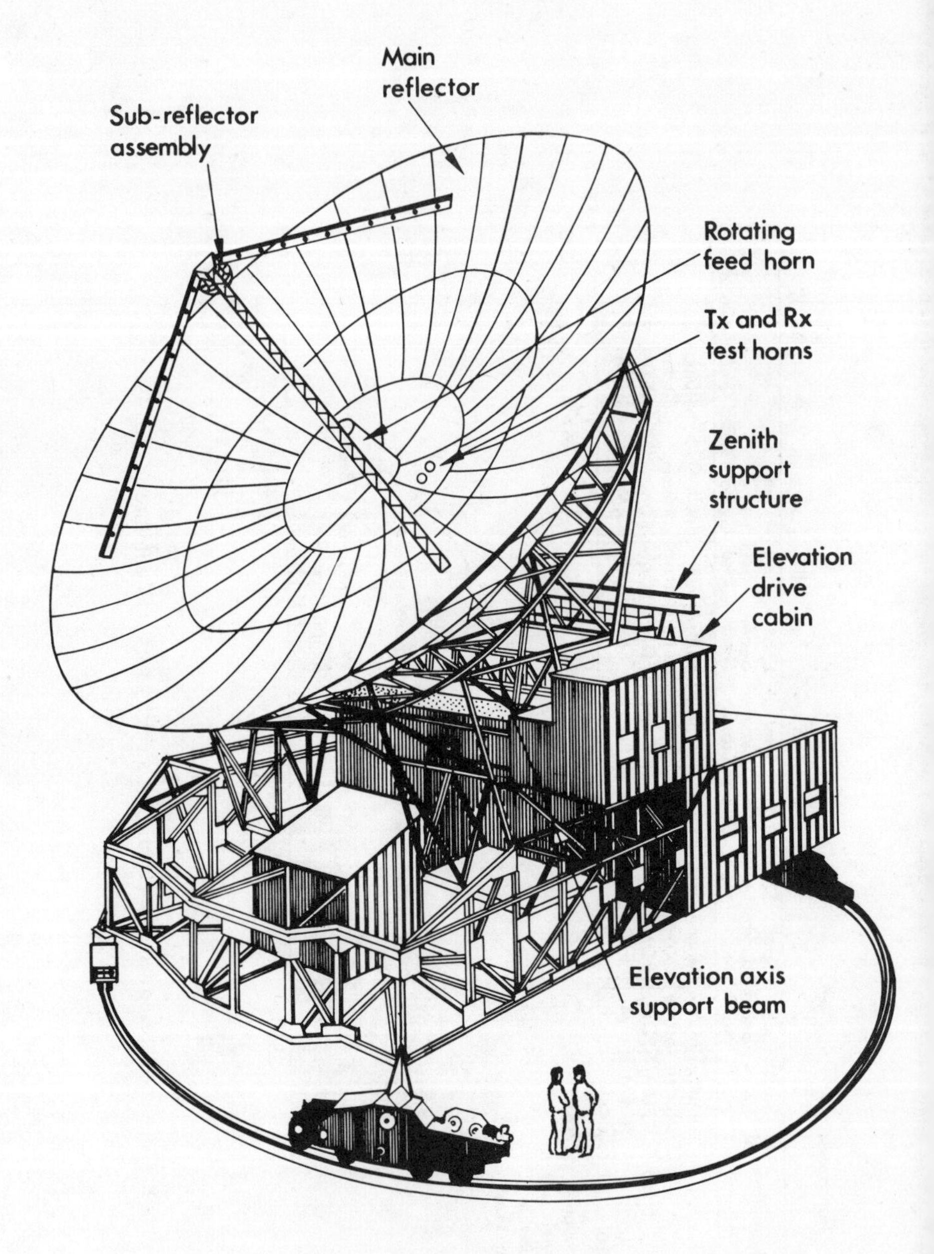

Figure 8.6 A typical cassegrain earth station antenna (Goonhilly II. U.K.). (Courtesy of British Post Office).

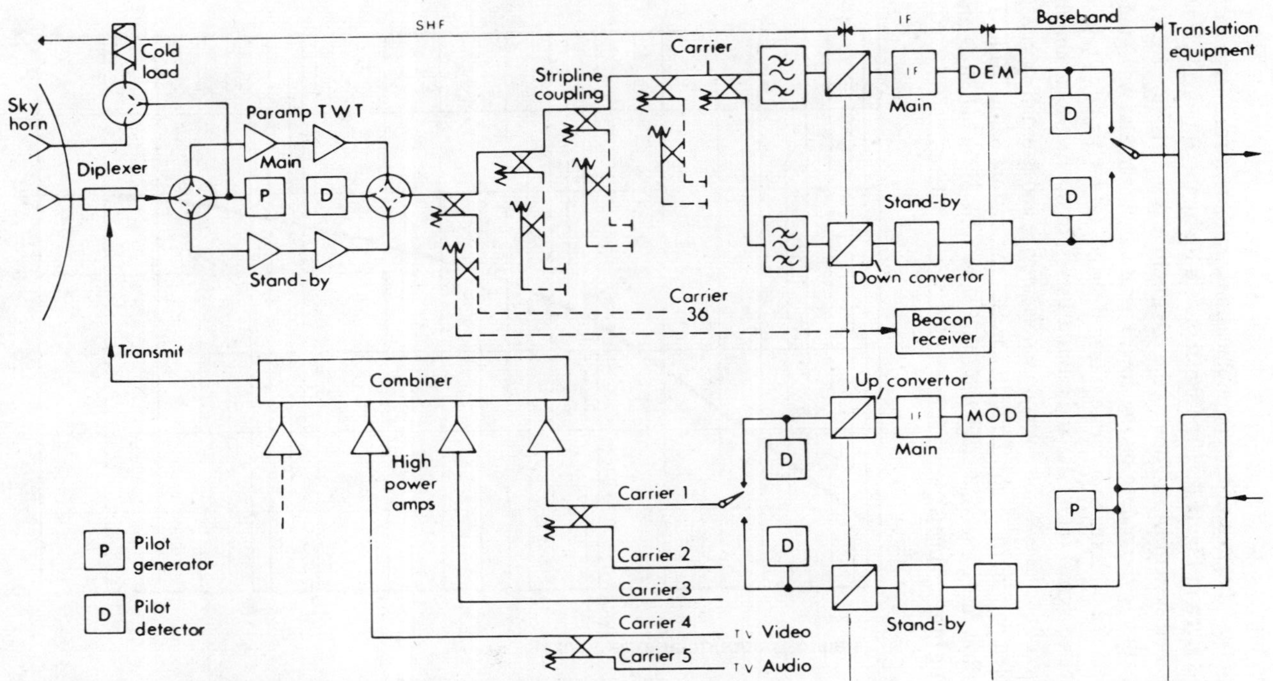

Figure 8.7 Outline diagram of modern earth station equipment

order of 26-27 m. The major types used are parabolic or Cassegrain and an example of these, which are equipped to automatically track the satellite, is shown in Figure 8.6.

An outline diagram of the communications equipment for a large earth station is given in Figure 8.7. The signal picked up by the antenna is fed via a low loss feed system which, due to the use of circular polarisation (one sense on the up-path and the reverse on the down), is usually in circular waveguide and incorporates a diplexer to separate transmit signals (at 6 GHz and up to 10 kW) from receive signals (at 4 GHz and around 1 pW), to the first stage amplifier. The latter has of necessity to be a very low noise device and is usually a cryogenically-cooled parametric amplifier with a noise temperature below 10 K. This is often followed by a second stage travelling wave tube or tunnel diode amplifier, branching into channels, down converted and demodulated.

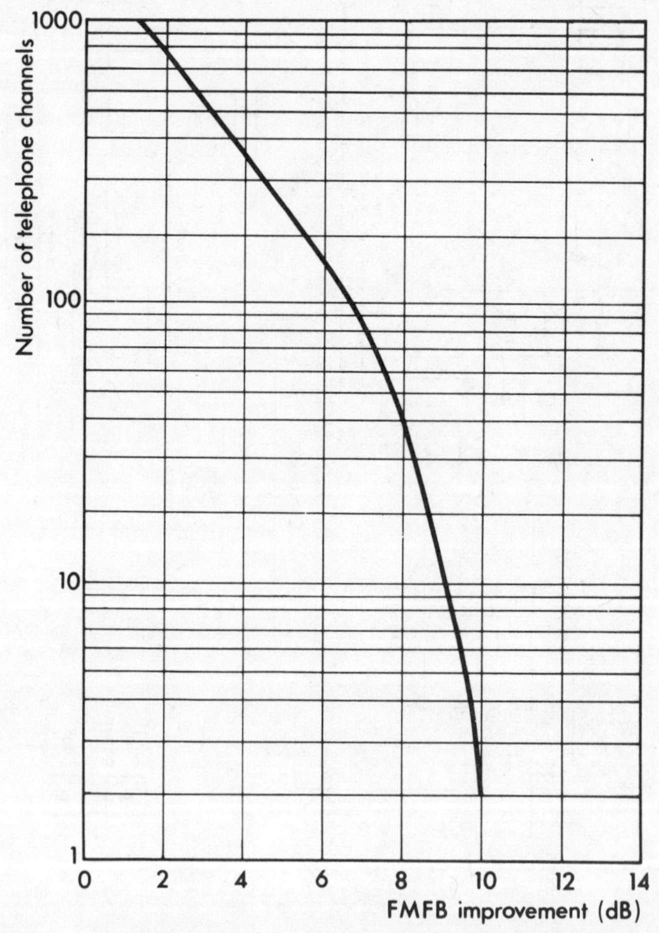

Figure 8.8 Threshold improvements due to F.m.f.b. Techniques

Conventional f.m. demodulators exhibit a threshold effect at around 10 dB as discussed in Chapter 7. Due to the high attenuation, earth stations operate close to the threshold and in periods of heavy attenuation may even drop below it. This would result in severe noise degradation with conventional devices, hence threshold demodulators [6] are used which effectively extend the threshold. The main types are (i) f.m. feedback, (ii) phase locked loop, or (iii) dynamic tracking, and typical extensions available for given channel capacities are shown in Figure 8.8. These devices may not be needed with the higher C/N obtained from type IV satellite spot beams.

On the transmitting side, the choice in both high power amplification and combination of the carriers lies between broadband and tuned devices. Broadband amplifiers such as the travelling wave tube are inefficient due to the back-off (see P 188) being applied, but with klystrons one per allocated carrier is required and one-for-one standby provided. Frequency-conscious channel-combining networks using circulators and filters may lead to difficulties if it is necessary to draw up a new frequency plan (because of failure of a satellite transponder, for example) since tunable high power filters are not available. But the alternative of broadband combiners using directional couplers is very lossy.

8.3 Systems equations

General noise power allowances are those of radio systems already discussed. The system is logically divided as follows:

(a) earth-satellite or up-link,
(b) satellite repeater,
(c) satellite-earth or down-link.

The segment of most concern is the down-link due to the limits of satellite output power and most noise is usually allocated to this. The maximum permissible 10 000 pW0p per telephone channel may be allocated as desired by the systems designer; the following is a breakdown recommended by INTELSAT:

Down-path thermal (N_D)	4250 pW0p
Satellite intermodulation (N_I)	2340 pW0p
Up-path thermal (N_U)	1410 pW0p
Transmitter earth station (all sources)	500 pW0p
Receiver earth station (excluding thermal)	500 pW0p
Interference with terrestrial links	1000 pW0p

(i) *Down-path* The down-path C/N is given as

$$(C/N)_D = E_S - L_D - M + G_R - N_D$$

where E_S is the satellite e.i.r.p. per carrier,
L_D is the free space plus atmospheric loss,
M is the operating margin,

G is the earth station gain (minus feed losses),
N_D is the down-path receiver noise, and is equal to
$-228{\cdot}6 + 10 \log_{10} (T_e) + 10 \log_{10} (B_{rf})$,
T_e is the earth station receiver noise temperature (see Appendix D),
B_{rf} is the occupied r.f. carrier bandwidth.

L_D obviously depends on the elevation of the earth station; most systems designs treat a worst elevation of 5° when the loss is 196.8 dB and atmospheric loss no more than 1 dB at this frequency.

M is due to signal variations due to heavy rain, variations in path length and transmitter power etc. and is usually taken as 4 dB. The system is thus operated at a C/N of 4 dB higher than theoretical threshold and the received signal may fall by this amount before threshold is reached.

N.B. The actual working C/N (C/N_T) is referred to an equivalent temperature at the receiver due to all the noise (up-path + intermodulation + downpath): thus a conversion must be made as follows:

$$\frac{C}{N_T} = \frac{C}{N_D} - 10 \log_{10} \left(\frac{10\,000 \text{ pW0p}}{\text{down-path thermal noise}} \right)$$

Incorporating the figure of merit into the down-path equation gives

$$\frac{C}{N_D} = E_S - L_D - M + 228{\cdot}6 + 10 \log_{10} (B_{rf}) + \frac{G}{T_e}$$

N.B. $\left(\frac{G}{T_e}\right)$ expressed in dB/K.

(ii) *Satellite repeater* The satellite repeater is characterised by saturating power amplifiers which due to the non-linear power characteristics give rise to intermodulation distortion and crosstalk. The need to obtain the most efficient use of primary power means operating the travelling wave tube near to saturation, and the non-linearity in this region is the cause of intermodulation between multiple carriers. In band intermodulation products may be caused by, (i) amplitude non-linearity causing third-order intermodulation terms, (ii) a.m. to p.m. conversion in the r.f. device causing the same, and (iii) gain variations accompanying the latter causing intelligible crosstalk.

In order to improve the C/N the repeater is operated further towards the linear part of the characteristics, and this results in output loss which is called back-off (b.o.). This is achieved by limiting the input power per carrier by controlling the earth station e.i.r.p. accurately. The optimum b.o. to obtain maximum capacity for the system is obtained by maximising the ratio $C/(N_D + N_I)$. A curve of C/N_D against b.o. may be plotted if (E_S – b.o.) replaces E_S in the down-link equation. Curves of C/N_I against b.o. have been produced by Westcott[7] and one is reproduced in Figure 8.9 for various numbers of carriers. Superposition of the two curves will readily yield the required back-off corresponding to the number of carriers in use.

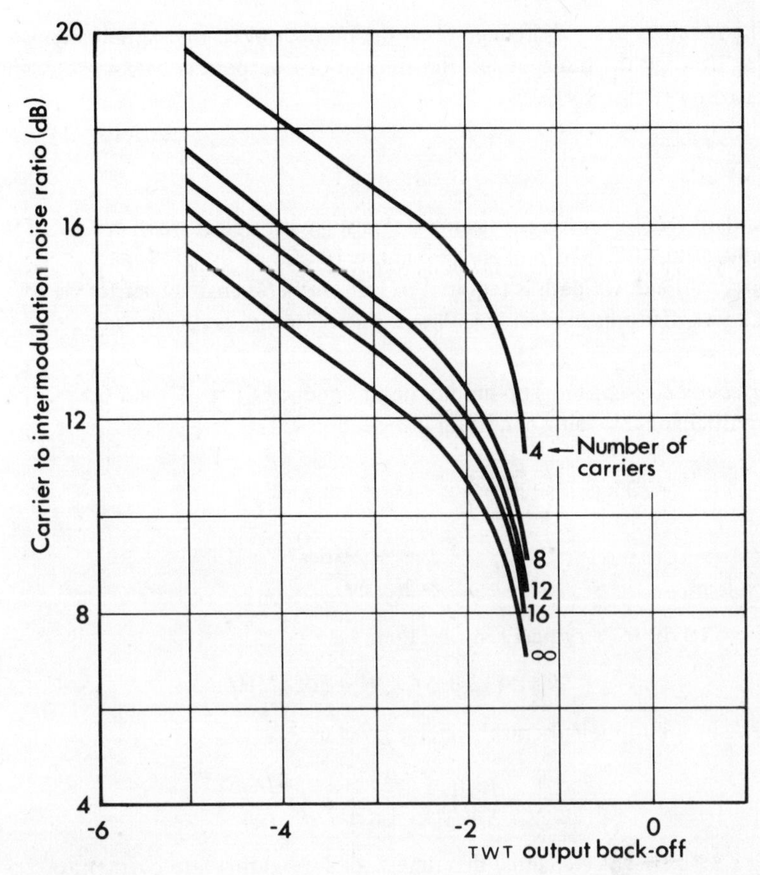

Figure 8.9 Carrier-to-total Intermodulation noise ratio as a function of back-off (from ref. 7)

(iii) *Up-path* The C/N at the input to the satellite first amplifier is given by:

$$C/N_U = E_e - L_U - M + G_S - N_U$$

where E_e is the earth station e.i.r.p.,

L_U is the up-path loss (200·3 at 6 GHz plus atmospheric losses),

M is the operating margin if required,

G_S is the satellite antenna gain,

N_U is the up-path thermal noise, equal to $10 \log_{10} (kT_S B_{rf})$,

T_S is the satellite system noise temperature.

The up-link C/N will be governed by the flux density received at the satellite which is controlled by the earth station e.i.r.p. as follows:

$$E_e = \text{power flux at satellite} + 10 \log_{10} (4\pi R^2) + \text{atmospheric losses}$$

where R is the slant height to the satellite (41 200 km for 5° elevation).

The required e.i.r.p. will depend on the number of carriers simultaneously passing through the transponder, the amount of transponder back-off etc. and is allocated by INTELSAT.

EXAMPLE

A standard INTELSAT earth station operating with a clear weather C/N of 9 dB and allocating 4000 pW0p of the maximum 10 000 pW0p allowed in a telephone channel to the down path is required to transmit a 60-channel carrier via an INTELSAT III transponder. Calculate the satellite and earth station e.i.r.p.'s.

(i) *Baseband calculation* Top modulating frequency $f_m = 12 + 4 \times 60 = 252$ kHz. The multichannel loading condition (see section 7.4) is

$$20 \log_{10} \left(\frac{F}{\Delta f_{\text{rms}}} \right) = -1 + 4 \log_{10} 60$$

and so $\quad F = 2{\cdot}03\ \Delta f_{\text{rms}} \quad$ (i.e. $l = 2{\cdot}03$)

Carson bandwidth $\quad B_{\text{rf}} = 2(cl\Delta f_{\text{rms}} + f_{\text{m}})$

Since $c = 10$ dB for a typical f.m.f.b., then

$$B_{rf} = (12{\cdot}8\ \Delta f_{\text{rms}} + 0{\cdot}505)\ \text{MHz}$$

The S/N in the top telephone channel is given as

$$\frac{S}{N} = \left(\frac{C}{N}\right) \left(\frac{B_{\text{rf}}}{3{\cdot}1\ \text{kHz}}\right) \left(\frac{\Delta f_{\text{rms}}}{f_m}\right)^2$$

taking a 3·1 kHz voice channel this time with a weighting improvement of 2·5 dB, where C/N is the operating value of 9 dB. Hence with Δ frms in MHz

$$S/N = (516 \times 10^3\ \Delta f_{\text{rms}} + 20{\cdot}35)\Delta f_{\text{rms}}^2$$

10 000 pW0p corresponds to a test tone S/N of 50 dB, or minus weighting and pre-emphasis 43·5 dB. Thus

$$2{\cdot}24 = (51{\cdot}6\,\Delta f_{\text{rms}} + 2{\cdot}035)\Delta f_{\text{rms}}^2$$

Solving, $\quad \Delta f_{\text{rms}} = 0{\cdot}266$ MHz

and so $\quad B_{\text{rf}} = 3{\cdot}904$ MHz

(ii) *Down-path calculation*

$$\frac{C}{N_T} = 9\ \text{dB} = 10 \log_{10} \left(\frac{C}{kT_T B}\right)$$

$$= \frac{C}{T_T} + 228{\cdot}6 - 10 \log_{10} (B_{\text{rf}})$$

Hence $$\frac{C}{T_T} = -123{\cdot}7 \text{ dBm/K}$$

$$\frac{C}{T_D} = -123{\cdot}7 + 10 \log_{10}\left(\frac{10\,000}{4000}\right) = -119{\cdot}7 \text{ dBm/K}$$

From the down-link equation,

$$E_S - \frac{C}{T_D} - \frac{G}{T} + L_D$$

(NB. The margin has already been added to the C/N threshold.) Thus

$$E_S = -119{\cdot}7 - 40{\cdot}7 + 197 = 36{\cdot}4 \text{ dBm}$$

Allowing for 0·3 dB atmospheric attenuation

Satellite e.i.r.p. = 36·7 dBm

(iii) *Up-path calculation* The type III specification is that $-40{\cdot}1$ dBm/m^2 produces a saturated output of 52·5 dBm.

Let us assume that for negligible intermodulation noise an input back-off 4 dB and output back-off of 2 dB is required. Hence an input flux density of $-44{\cdot}1$ dBm/m^2 produces an output of 50·5 dBm.

Actual 60-channel output e.i.r.p. = 50·5 – 36·7
= 13·8 dB down on saturated value

Thus input flux density required = –44·1 – 13·8
= $-57{\cdot}9$ dBm/m^2

From the up-path equation,

$$E_e = \text{power flux} + 10 \log_{10}(4\pi R^2) + \text{losses}$$
$$= -57{\cdot}9 + 163{\cdot}3 + 0{\cdot}3 = 105{\cdot}7 \text{ dBm}$$

assuming a 5° earth station elevation.

8.4 Digital systems

In keeping with a general resurgence of interest in digital systems brought about by improved hardware and the increased capacity of heavy duty routes, digital satellite communication is the logical extension from the f.m.-f.d.m.a. system. The latter offers a potentially greatet flexibility in multiple access techniques which would be accompanied by a reduction in earth and space segment costs. Another important consideration is that up to the present satellites have been power limited, but in the future this will change to bandwidth limitation where the digital techniques are almost certainly more beneficial. But the aims for the future must also be biassed towards incorporating more users into the global system, possibly extending the benefits to smaller countries whose international traffic does not at present warrant the smallest pre-assigned INTELSAT block rental.

With these aims in mind, two new systems have been developed, and whilst they may not in detail form the basis of the new system, there can be no doubt that their basic configuration will.

A demand assignment system

The busiest INTELSAT routes carry several supergroups, but there are many routes which carry only a handful of channels and many with only one. f.m.-f.d.m.a. transmission is efficient for large basebands but much less efficient for small ones split up between many carriers. There is thus a need for an alternative to the present system designed specifically to serve small routes efficiently.

There have been several systems proposed [8, 9], but the one chosen by INTELSAT for further development is called SPADE, which is a p.c.m.-f.d.m.a. demand assigned system. As already indicated in Figure 8.5, a decision has been taken to use this system in one of the 12 transponders of the type IV satellite. The 36 MHz bandwidth of the transponder will be split into 400 pairs of radio channels, each of 45 kHz width. Each pair of these frequency slots is used for a single return telephone circuit. Any idle pair can be seized by two earth stations with a call to connect, the pair being replaced in the common pool as soon as the caller hangs up. A diagram of the extra earth station equipment is shown in Figure 8.10. Speech signals from the International Exchange are converted from analogue to 7-bit p.c.m. at the transmitting earth station, then modulate the carrier frequency appropriate to the selected slot in 4-phase phase-shift keying. The carrier is switched off to save satellite power when the speaker is silent. The process is reversed at the receiving earth station, and an analogue signal is passed on to the distant International Exchange. Control of the seizure of frequency slot pairs, the transmission of channel signalling information and various other ancillary functions are exercised by automatic data processing units at each earth station, linked together by a common signalling time division multiple access data channel which is also transmitted through the satellite system.

The important innovation in the SPADE system is the facility by which an earth station occupies capacity in the satellite system only while a call is in progress. This is called 'demand assignment'. In exchange for slightly more complicated earth station equipment, it offers an efficient way of transmitting calls on lightly loaded routes, thus opening the door to the small routes and increasing satellite bandwidth utilisation, thus reducing space segment costs.

The combination of f.m.-f.d.m.a. and a demand assigned system will form the nucleus of the new system. Whilst pre-assignment still remains, the best choice for heavy duty routes and f.m.-f.d.m.a., the best transmission means the optimum transmission method for the demand assigned, or single channel per carrier system is by no means so well defined. Many small nations will still be

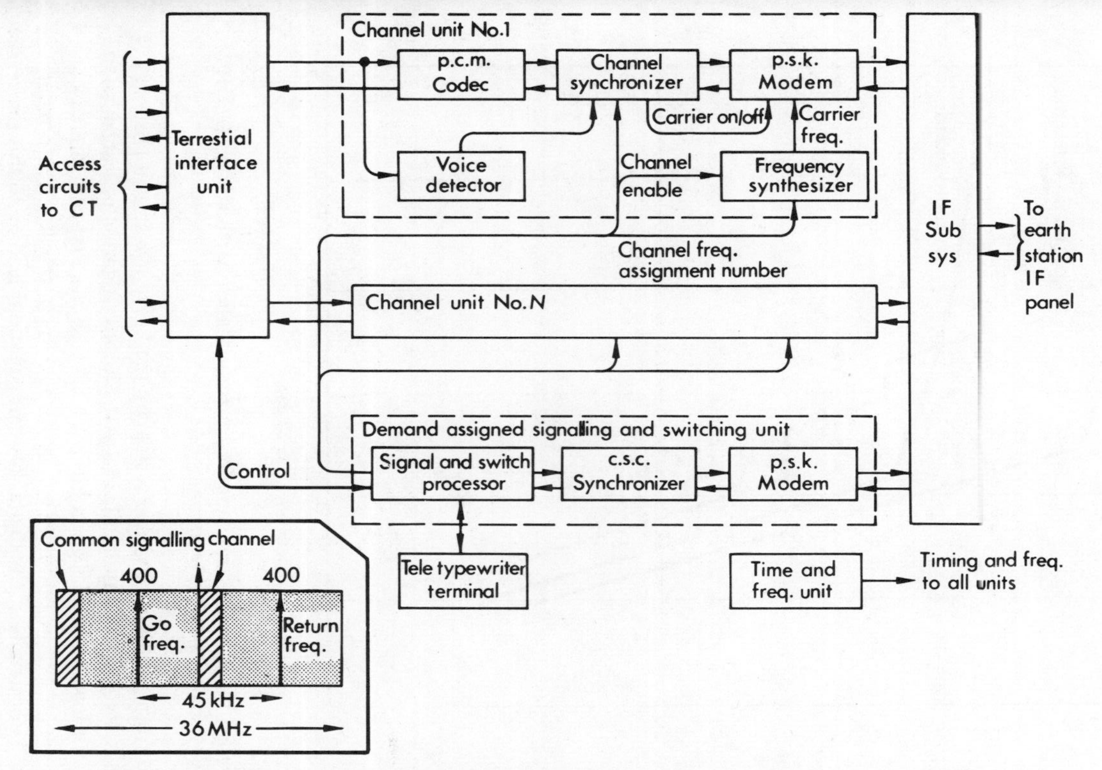

Figure 8.10 Simplified block diagram of earth station equipment for the SPADE system showing insert SPADE transponder frequency plan (from ref. 8)

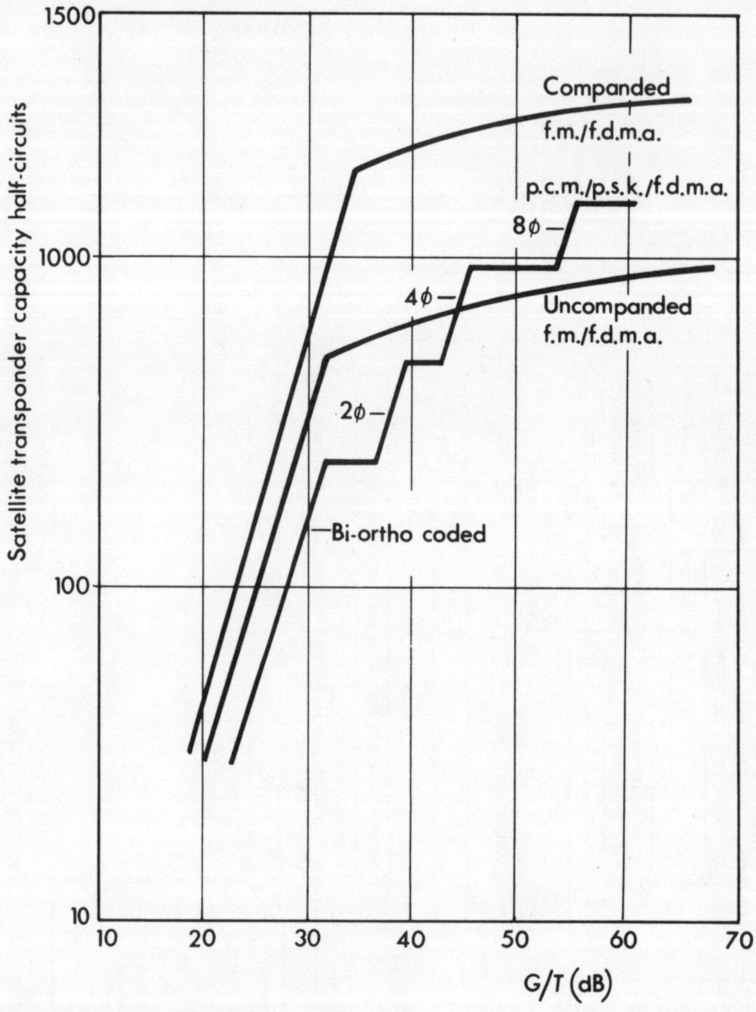

Figure 8.11 Satellite transponder capacity against earth station G/T for various transmission systems

deterred from entering the global system by the heavy investment costs of a 40·7 dB/K station still required for SPADE. An investigation of transmission systems for lower G/T's, as given in Figure 8.11, shows that companded f.m.-f.d.m.a. is a more optimum choice for maximum satellite transponder capacity. (For details of the digital systems equations leading to this figure, see Appendix E). This system, provided that compandors can be made to operate effectively in this mode, could well form the basis of rival systems which may take the form of domestic or regional systems. The introduction of small stations ($<$40·7 dB/K G/T) is inevitable but INTELSAT policy regarding their inclusion in a mixed system has so far not been formulated.

A time division multiple access system
The use of f.d.m.a. for transmitting groups of multiplexed channels has two basic disadvantages; firstly, that to avoid excessive intermodulation from the multiple carriers through the satellite transponder the output power of the satellites requires limiting, and secondly, the inflexibility of the system in that to change the baseband assembly requires modification of the receive and transmit earth station equipment and leads to problems of frequency planning. The use of t.d.m.a. avoids these difficulties as each earth station in turn transmits a burst which occupies all of the frequency spectrum of a transponder, and since only one carrier is being transmitted at any time, no complicated frequency plans are required and the amplifiers can be run at their saturated power levels. Since each earth station transmits for only a fraction of the available time, t.d.m.a. is used with digital rather than analogue transmission methods.

Figure 8.12 shows a sequence of transmissions (p.c.m.-p.s.k.-t.d.m.a.) from earth stations; and the more detailed structure of the transmission or 'burst' associated with a single station is shown in Figure 8.13[10]. The speech or data information contained in each burst is preceded by a pre-amble which consists of the following:

(a) Guard period to avoid overlap of transmission due to timing errors.

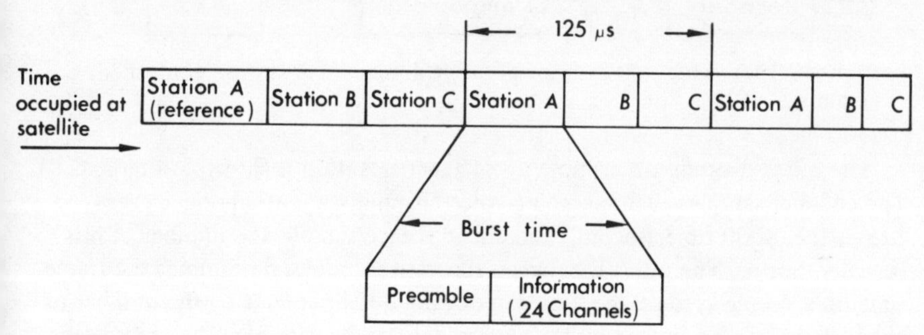

Figure 8.12 Frame structure of simple 3-station t.d.m.a. system (from ref. 10)

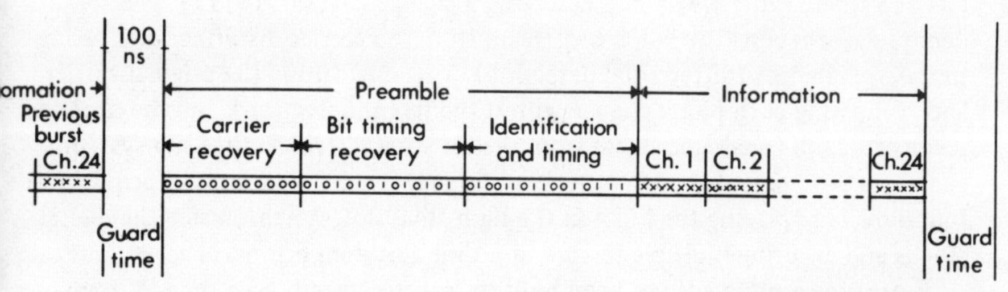

Figure 8.13 Burst structure for 24-channel t.d.m.a. system (from ref. 10)

(b) A period of unmodulated carrier to enable the demodulator to generate a coherent carrier for detection (carrier recovery).
(c) A number of bits which ensure that the clock associated with the digital equipment at the receiving station is running in the right phase (synchronisation).
(d) A unique word, i.e. a pattern of digits unique to the transmitting station which serves to identify it and enables the various bursts to be correctly related in time.

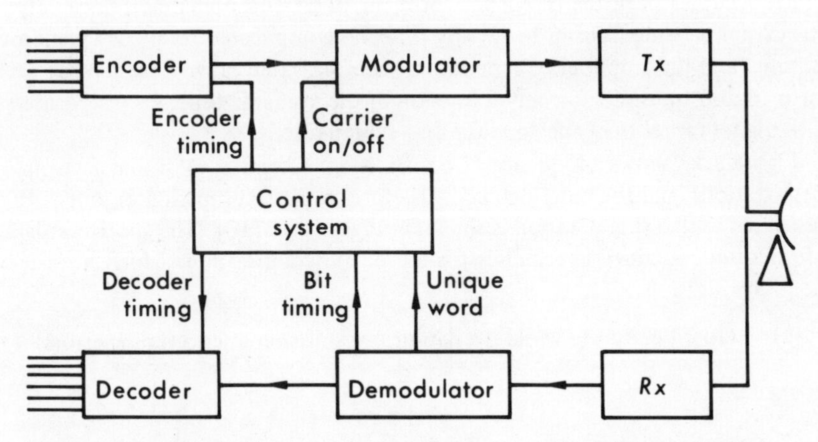

Figure 8.14 Block diagram of p.c.m./t.d.m.a. earth station equipment (from ref. 10)

The block diagram of a simple t.d.m.a. earth station is shown in Figure 8.14. The encoder samples each incoming voice channel sequentially once every 125 μs (i.e. 8000 times/second) and encodes each sample as a number of bits (usually 7 or 8). The interval between successive samples determines the frame time on a simple system; the duration of each burst depends on the number of channels to be transmitted by the station and can be variable. The modulation system used is 4-phase p.s.k. which gives the maximum bandwidth to power trade-off and a shift of the carrier phase by 90 degrees is made by looking at two bits at a time in the stream and detecting a change (01, 00, 10, 11). The demodulator reconstitutes the bit stream from the carrier, by either comparing successive changes (differential detection) or reconstituting the original carrier and comparing each phase change with it (coherent detection), and the decoder reconstructs the analogue signal from the bit stream. The control sub-system generates the pre-amble and controls all timing, synchronisation and acquisition functions (e.g. placing the burst in the right time slot, synchronising the digital clocks and ensuring multiple seizure of a time slot does not occur).

Prototype equipment has been built to test the system via a type IV transponder. Further economies in power and bandwidth could be obtained by use of higher phase p.s.k. for the high power spot beams, (note the improvements, Figure 8.11), or by encoding and transmitting samples from each channel whilst they are active, or both. The latter method would combine the advantages of demand assignment and TASI at the cost of some considerable complexity though. It is envisaged that this very flexible system would take over from the f.d.m.a. system when capacity and equipment availability warranted it.

8.5 Domestic systems

Domestic services via satellite are becoming an economic alternative to long terrestrial systems using cable or radio-relay for the larger countries, and the break-even distance will shorten as time passes.

The form of the system would depend very much on the ecology of the country concerned; Canada for example is planning a domestic service[11] largely to distribute television to the remote areas such as Alaska. This is the sort of system already employed in the USSR. In large countries where the population is thinly spread a simple form of satellite communications may be cheaper than terrestrial links. For instance, Australia is studying a simple delta modulation system[12] to replace the h.f. radio telephone network that serves the 'outback'. A system is also under way to distribute television to large community receiving systems in India using f.m. and carrier frequencies of 850 MHz with 7 m antennas and simple receivers.

Undoubtedly the major systems will be those to serve large areas with large traffic densities where links between a small number of provincial centres will be used with large earth stations. The U.S[13] and European areas fall into this class and already plans are being formulated for their domestic systems. Due to the high power satellites employed to cope with the large predicted traffic increases these latter systems will almost definitely look to the higher frequency bands (above 10 GHz where atmospheric attenuation may be a problem) and be digital (possibly p.c.m.-p.s.k.-t.d.m.a.). The planning of these systems involves studying future traffic matrices and optimising the earth station G/T for the system. For example, as G/T is decreased earth sector costs per circuit decrease, but since satellite capacity falls the space sector costs per circuit rise, so there is an optimum corresponding to minimum costs. There are almost certain to be constraints on the bandwidth and methods of increasing capacity relative to that of the basic system will be used. Possibilities for the latter are:

(a) Increased transponder powers. This is an obvious way of increasing capacity but relies on state-of-the-art in equipment and achievable energy dispersal to minimise interference.

(b) Use of spot beams and multi-level p.s.k. with speech interpolation. Spot beams cut down the area coverage and require higher phase p.s.k. equipment (not yet available) to exploit them in the t.d.m.a. mode. Unless on-board switching is provided so that each transponder can be connected to any beam, the use of spot beams leads to reduced flexibility. Also spot beams on the up path could lead to interference between satellites and place severe restrictions on the capacity of the geostationary orbit.

(c) Frequency re-use. By using the discrimination between two orthogonal planes of polarisation on both up and down-links the capacity is doubled. Problems here are the effects of precipitation on the discrimination and the dual polar feeds and diplexers required in the earth stations. If the latter can be solved this looks a very attractive way of increasing capacity.

Arguments for and against regional satellite systems tend to be based on economics and break-even distances with terrestrial systems, but the true advantage of satellite systems lies in their flexibility to deal with changing traffic conditions, and it is this aspect which will enable them to be a serious transmission competitor in the future.

REFERENCES

1. A. Clarke, 'Extra Terrestrial Relays', *Wireless World*, October 1945, pp. 305-308.
2. R. Gould and G. Helder, 'Transmission Delay and Echo Suppression', *I.E.E.E. Spectrum*, April 1970, pp. 47-54.
3. F. Taylor, 'INTELSAT–the International Telecommunications Satellite Consortium', *I.E.E. Electronics and Power*, January 1971, pp. 8-13.
4. J. K. Jowett, *Technical Arrangements for the Global System*, United Kingdom Seminar on Satellite Communications, May 1968, Paper A2.
5. *ICSC Documents* – 'ICSC-37-38E-Performance Characteristics of Earth Stations'.
   CCIR XIIth Plenary Assembly, New Delhi 1970, Vol. IV, Part 2, Fixed Services Using Communication Satellites'.
6. C. M. Thomas, '*Principles of threshold extension demodulator operation*' communications satellite earth station technology, Washington 1966.
7. R. Westcott, 'Investigation of Multiple f.m./f.d.m. Carrier Through Satellite TWT Operating Near to Saturation', *Proc. I.E.E.* **114**, 6, June 1967, pp. 726-740.
8. A. Werth, *Spade: A p.c.m. f.d.m.a. Demand Assignment System for Satellite Communications,* International Conference on Digital Satellite Communications, London, 1969, pp. 51-58.
9. B. Evans and R. Walters, *An Economic Satellite Communications System for Small Nations*, IEEE International Communications Conference, 1971, Paper 19D, pp. 1921-1925.
10. W. Schmidt, *et al., MAT-1 –INTELSAT's Experimental 700 Channel t.d.m.a./d.a. System*, International Conference on Digital Satellite Communications, London, November 1969, pp. 428-440.
11. J. Almond and R. Lester, *Communication Capability of the Canadian Domestic Satellite System*, IEEE International Conference on Communications, June 1971, Paper 11A, pp. 11.1-11.7.
12. D. Snowden, *A Small Station Satellite System using Delta Modulation*, International Conference on Digital Satellite Communication, London, 1969, p. 26-38.
13. L. Tillotson, 'A Model for a Domestic Satellite Communications System', *B.S.T.J.* December 1968, pp. 2111-2137.

Chapter 9

Programme and television transmission

9.1 General

In addition to telephony signals, there is also the need in a telecommunication system to transmit video and high-quality audio signals. The main users of these services are the radio and television broadcasters who need to connect their programme sources to a control centre (*contribution network*) and the control centre to transmitters (*distribution network*). These services are provided in the main by the telecommunications authority of a country, and share physical facilities such as ducts and microwave towers with the telephony system. In addition wide-band f.d.m. systems are used to carry television signals instead of multi-channel telephony provided they have been designed with this possibility in mind.

Another area which is becoming of increasing importance is television distribution to subscribers. This may be for domestic television where radio reception is unsatisfactory or uneconomic, so-called Community Antenna Television (CATV) or for educational use for schools or colleges. There is also the possible need for a lower quality television for use with viewphone systems.

The detailed technology of programme and video transmission is beyond the scope of this book and this chapter discusses only those aspects which are relevant to transmission systems designed to carry these other services as well as or alternatively to multi-channel telephony.

9.2 Programme transmission

The objective of broadcasting is to provide as high a quality of transmission as possible, and this implies that the transmission system design is based on all the perceivable imperfections being negligible. The characteristics of signals produced by studios contain significant energy in the band 40 Hz to 15 kHz and a dynamic range of up to 70 dB for orchestral music[1]. From practical tests the level of allowable degradations are[2]:

(a) *Frequency response* – ±0·5 dB from 125 Hz to 10 kHz and not more than –2 dB at 40 Hz and 15 kHz.

(b) *Noise power.* Noise measured with appropriate weighting network should be more than 60 dB less than the peak power.
(c) *Non-linear distortion* should be less than 1 per cent
(d) *Error in reconstructed frequency* should be less than 1 Hz.

In addition, if a pair of channels are used for stereophonic transmission then the level difference between the two channels must not exceed 0·8 dB and the phase difference must not exceed 15°-30°.

These specifications are very severe and frequently circuits of lower quality have to be used. In particular for international use, circuits with bandwidths 50 Hz to 6·4 kHz or 50 Hz to 10 kHz are used. These are designed to occupy two or three 4 kHz channels in an f.d.m. group.

In a one-way transmission system with amplifiers it is necessary to specify what level the signal is to be transmitted. This is done by specifying the quasi-peak power as measured on a suitable volume meter such as described in Chapter 1. For international use the sending level is adjusted so that the quasi-peak power is +9 dBm0. The actual peak power will depend upon the characteristics of the type of programme and it is found that it rarely exceeds +15 dBm0. The mean power will again depend upon the type of programme and measured values range from −9 to −3 dBm0 for a one-minute integration period. These figures may be changed by the fact that frequently signals have been submitted to selected emphasis in the studio for artistic requirements.

The general characteristics for a programme transmission circuit may be found by considering a hypothetical reference circuit consisting of a chain of circuits in a similar manner to that described in Chapter 3 [3]. It is found that when programmes have to be transmitted on an f.d.m. system the noise performance is usually inadequate and a compandor has to be used.

9.3 Television transmission

Signal characteristics
Broadcast television signals nearly all fall into one of two classes:

(a) 625-line 50-frame standard requiring 5 or 5·5 MHz of video bandwidth.
(b) 525-line 60-frame standard requiring 4·2 MHz of video bandwidth.

Where colour is provided then in addition to the luminance (or black and white) component, it is necessary to send two additional signals which give the hue (i.e. which colour) and saturation (i.e. how strong) of the colour. There are three main coding techniques used in broadcast television for sending this chrominance information and they are all designed to utilise the same bandwidth as the monochrome signal and to provide a compatible picture which may still be received on a monochrome receiver. All the techniques do this by using a sub-carrier near the top of the luminance band. The exact frequency is chosen to be an odd multiple of half the line frequency, which ensures that the components of the chrominance signal spectrum is interleaved between the luminance spectrum, and it is this which ensures compatibility. The colour sub-carrier frequencies used are at 4·43 MHz for 625-line systems and at 3·58 MHz for 525-line systems.

The three techniques are:

(a) *NTSC* (*National Television System Committee*). This is the original system as used in North America and uses the amplitude of the sub-carrier to send the saturation and its phase to send the hue.

(b) *SECAM* (*séquential couleur à memoire*). This system is an improvement on the NTSC system and has been developed in France. With this system the two chrominance signals are transmitted alternately during successive scanning lines using frequency-modulation of the sub-carrier. The receiver needs a delay line in order to present the two signals simultaneously.

(c) *PAL* (*phase alternation line*). This was invented in Germany and overcomes one of the principle defects of the NTSC system (its susceptibility to phase errors) by reversing the polarity of one of the two chrominance components every alternate line. The two signals are transmitted as amplitude-modulated signals on a pair of suppressed sub-carriers in quadrature at the same frequency.

Since systems (a) and (c) need phase information, it is necessary to place a burst of sub-carrier during the synchronisation pulse preceding each line in order to lock the local oscillator. In system (b) it is necessary to provide line identification.

The relative merits of these systems and yet further proposed improvements are beyond the scope of this book. What is important is the requirements placed on the transmission systems used for the contribution and distribution networks of the broadcasters. These networks almost invariably use the same signal format as is actually broadcast, i.e. there is no separate transmission of the chrominance signals.

Channel characteristics

A wide-band channel provided for f.d.m. use is designed to have a very flat frequency response (usually better than ±0·5 dB overall). However, this is not necessarily sufficient for a good television channel since it is the time response of a channel which is important as well. The main time impairments are ringing and echos. Ringing is produced by an amplitude/frequency response with a finite pass-band with a sharp cut-off at the top of the frequency band. An echo may, of course, be produced by a reflection at an impedance discontinuity, but the frequency response of a system with a single echo is one which has small but regular variations. Hence conversely a system which has small but regular variations in its frequency response will produce an echo. This is most likely to occur in practice on a coaxial circuit. The time response may be corrected by

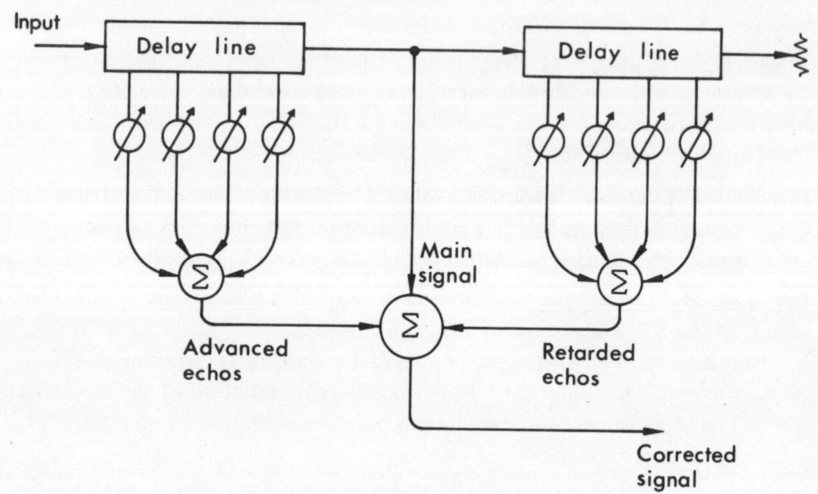

Figure 9.1 A transversal filter for video signal correction

means of what is called a transversal filter[4] as shown in Figure 9.1. With the aid of this equipment it is possible to cancel out the echos due to reflections and regular frequency response variations.

In a colour signal, since the luminance and the two chrominance signals are sharing the same bandwidth then the effects of non-linearities will cause intermodulation, and this may produce peculiar colour effects. Hence strict limits are placed on the differential-gain and differential-phase of a channel, i.e. the difference between the small-signal gain and phase response at the sub-carrier frequency when superimposed on a luminance signal corresponding to black and one corresponding to white.

Finally the effect of noise, thermal, cross-talk and power supply hum in the case of a base-band circuit has to be considered.

Channel Objectives

The overall channel objectives must be found by practical experiment using many subjects, and the same difficulties are found as have been mentioned for telephone systems. These are to find a suitable common factor against which to measure all the various impairments and then to find appropriate additional laws for the individual impairments. For monochrome systems a useful common factor which has been chosen as a result of experience is an exact echo $\frac{4}{3}\mu s$ after the main signal and of adjustable amplitude. The relative amplitude of this echo is called the *K-rating* of the system and is usually expressed as a percentage[5]. It is found that for a good overall system a rating of less than 3 per cent is desirable and that individual contributions add on an r.m.s. basis. Hence an individual link should have a rating of much less than 3 per cent. It is now possible to judge each impairment against the *K*-rating to get an overall figure of merit.

In North America a similar scheme is used [6] to give an *echo rating* whereby individual echoes are weighted according to their time of occurrence and then summed on an r.m.s. basis.

For colour transmission the problem of assessing the tolerable limits of channel impairments is very great as it will depend upon the coding system and the effects are more difficult to quantify [7]. Typical results are that the differential gain should not be more than 10 per cent and the differential phase more than 5° for an overall system.

Having obtained a set of overall objectives, the objectives for an individual channel may be obtained by constructing suitable hypothetical reference connections [8]. In a very long system it is possible to provide adaptive circuits which can reduce the differential gain and phase distortions of a circuit. These operate on special test signals inserted in the fly-back lines of the signal.

Use of f.d.m. transmission systems

6 MHz f.d.m. coaxial systems are usually designed so that they may carry a 5·5 MHz television signal as an alternative. This may be achieved by a.m. vestigial side-band modulation for the television signal and suitable adjustment of the sending level in order to make its peak power acceptable to the repeaters. Some pre-emphasis is usually used to improve the noise performance. It is not usually possible to send the sound signal on a separate carrier in the same band as the necessary reduction of the power of the video and audio signal to prevent intermodulation will usually increase the noise on the video signal to an unacceptable level. The sound signal must therefore be sent by a separate channel, which introduces complications. The problem may be overcome by coding the sound signal in p.c.m. and placing the code in the line blocking portions of the signal [9] (two samples per line giving a practical audio bandwidth of 14 kHz). This does not introduce any additional loading effects on the repeaters and it is planned to use this technique extensively in the BBC television network.

9.4 Cable television

There is an increasing use, especially in North America, of distribution of television to the home by the use of cables [10]. The reasons for this are several:

(a) lack of adequate signal strength from broadcast station,
(b) increase in number of channels made available,
(c) specialised services such as educational use, distribution of stock market prices etc.

The frequency bands chosen for distribution depend upon the design of the receivers, and the actual channels are chosen to ensure that there is a low likelihood of interference with broadcast stations. When several channels are sent along a single cable then the amplifiers must be designed to produce negligible intermodulations. For instance, in London there is a schools' distribution network which transmits a total of 9 channels to 1200 schools on a single cable in the v.h.f. band 40 to 140 MHz[11].

REFERENCES

1. 'Characteristics of Signals Sent Over Sound Programme Circuits', CCIR Report 491, XII Plenary Assembly, Volume V, 1970.
2. 'Circuits for High-Quality Monophonic and Stereophonic Transmission'., CCIR Report 496, XII Plenary Assembly, Volume V, 1970.
3. *C.C.I.T.T. White Book*, Volume III, Recommendation J21.
4. J. M. Linke, 'Some Aspects of Echo Waveform Correction and Synthesis', *Proc. I.E.E.* **110C**, 1963, pp. 213-222.
5. I. F. MacDiarmid, 'Waveform Distortion in Television Links', *Post Office Electrical Engineers Journal* 52, 1959, p. 108.
6. *Transmission Systems for Communication*, Western Electric, 1970, Chapter 29.
7. L. E. Weaver, 'The Quality Rating of Color Television Pictures', *Journal of Society of Motion Picture and Television Engineers* 77, 1969, pp. 610-612.
8. *C.C.I.R. Recommendations* 421-2 and 451-1 (XII Plenary Assembly 1970).
9. J. M. Chorley, and D. E. C. Shorter, 'P.c.m. "sound-in-syncs": Operational Systems for Video Distribution and Contribution Networks', International Broadcasting Convention 1970, *I.E.E. Conference Publication No. 69*, pp. 166-169.
10. For more details, especially of the commercial problems involved see the I.E.E.E. Special Issue on Cable Television, *Proc. I.E.E.E.* 58, July 1970.
11. J. E. Haworth, 'The School-Television Distribution Network', *Post Office Electrical Engineers Journal* 61, January 1969, pp. 245-250.

Chapter 10

P.c.m. systems

10.1 Basic principles

The basic principles of p.c.m. systems are well described in many text-books[1] and the basic components of a system are summarised below:

(a) *Sampling circuit.* The signal is sampled at slightly more than twice the highest frequency so as to permit the use of fairly simple band-limiting filters. The sample produced will have a finite width and a non-flat top. Since the encoding process takes a finite time it is usually necessary to have a hold circuit which will produce a constant voltage equal to the sample value for a sufficient time to allow encoding. If the width of the sample is such that the top is significantly non-flat, then the averaging effect produced by the hold circuit will introduce frequency distortion.

(b) *Encoder.* The encoding process finds a quantised level which is nearest to that of the signal and this level can then be described by a suitable code, such as binary. The process of quantisation introduces noise, since a signal reconstructed from the quantised levels will differ from the original signal. If the quantal levels are equally spaced in the amplitude range, then the noise will be proportional to the difference in quantal levels and will be independent of signal level (provided a signal is present). For speech signals, it is found that most of the information is contained in the waveform around the axis, whereas a degree of limiting of the waveform will have no noticeable effect. Also it is found that, subjectively, noise is less objectionable for high amplitude signals than for low amplitude ones. Hence, it is desirable to have some form of non-linear quantisation.

It may be shown that if the quantum sizes are approximately proportional to the logarithm of the signal level then it is possible to achieve an approximately constant signal-to-quantisation-error power over a wide range of speech volumes[2]. A truely logarithmic characteristic would obviously require an infinite number of steps, so it is usual to use a characteristic which is logarithmic at large amplitudes and linear for low amplitudes. The exact characteristic will depend upon the range of speech characteristics over which the system must work[3]. The effect of companding is to extend the range over which a minimum signal to noise ratio is obtained. The amount of this reduction,

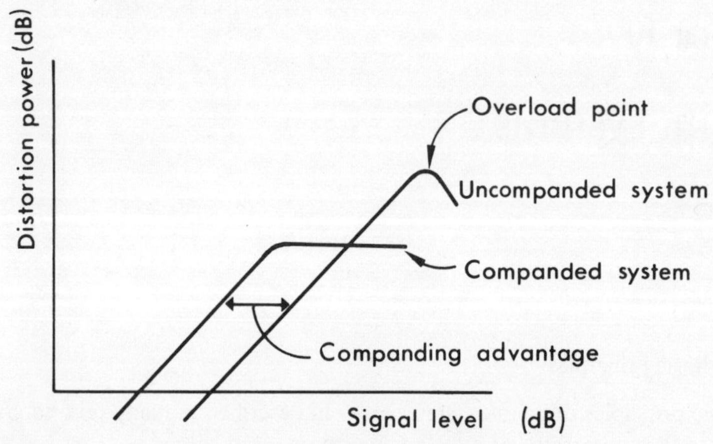

Figure 10.1 Definition of companding advantage

measured in dB, is referred to as the *companding advantage*. This is shown in Figure 10.1.

In practical systems the signal compression may be achieved in several ways. The simplest is to pass the signal through a non-linear circuit with a suitable transfer characteristic which compresses the signal and follows it by a uniform encoder. At the remote end there is a complementary expansion circuit after the decoder. Obviously it is necessary to match accurately the characteristics of the units to avoid overall non-linear distortion. Alternatively, a uniform encoder may be used, followed by logical circuits to produce a suitable code. This latter requires a more sensitive encoder but the circuitry is not so subject to temperature variations as is the former case. A third alternative exists whereby a non-linear encoder can be used directly.

The effect of varying the quantum size is that of an instantaneous compandor; since there is exact reproduction of the digital signal at the remote end, then there is no need to use a syllabic compandor.

(c) *Transmission* [4]. The code giving the quantal level is converted into a form suitable for transmission; a common choice is some form of bipolar code as this makes for the simplest detection at intermediate points. Since the transmitted code can have only one of a finite number of values then it is possible to use regenerators rather than repeaters and these can work at a considerably worse signal-to-noise ratio than a repeater. For instance, a repeater for an analogue signal needs to be used wherever the signal-to-noise ratio at the worst frequency falls to about 60 dB, whereas if it is only necessary to detect whether a pulse is present or not then a signal-to-noise ratio of only about 20 dB will be adequate. The transmitted code is not limited to binary and multi-level codes and regenerators are practicable in some circumstances (see later).

This is the crucial advantage of p.c.m., that the use of regenerators allows virtually distortionless transmission of a digital signal, irrespective of distance or routing, whereas in an analogue transmission system, the impairments are cumulative. In a p.c.m. system the most significant impairment is the quantisation noise and this will be constant for a given digital system. Also, in a p.c.m. system the gain of the circuit depends only upon the accuracy of the coder and decoder and is not dependent upon the transmission medium. Hence a p.c.m. link may be operated at a fixed loss (usually 3-6 dB) irrespective of the number of digital sections in tandem.

(d) *Pulse shape.* For simple binary pulses the waveform to the line is usually rectangular. The frequency characteristic of the cable will introduce dispersions and the effect of this is to spread the received pulse out so that it occupies more than one time slot. This produces inter-symbol interference and reduces the reliability of detection. The loss of the cable will introduce thermal noise, and in a practical system there will be crosstalk between physically adjacent cables. For these reasons it is necessary to use an equaliser prior to detection in order to minimise the effect of inter-symbol interference and the effects of noise and crosstalk. The design of these equalisers is complex and usually relies on a computer optimisation procedure.

It is necessary to provide a local timing waveform in a regenerator and this may be obtained by the use of a high Q (about 1000) resonant circuit at the bit frequency of the input stream. Because of the random nature of the transmitted bit stream there will be some random phase variations of the timing waveform, and this effect will be cumulative in a long system[5]. The overall effect will be to cause a time jitter of the pulses at the end of the system which will give rise to noise due to pulse-position-modulation. This is an accumulated impairment since it increases with the number of regenerators in tandem. However, the amount of jitter may be reduced on a long line by means of a circuit which reads the pulse stream cyclically into a store; the information may then be read out at a uniform rate derived by a very high Q circuit operating on the incoming bit stream (see Figure 10.2).

(e) *Reconstruction.* At the receiver the digits are decoded into quantised levels and passed through an expansion circuit (or alternatively the received digits are operated upon digitally and a uniform decoder is used).

The power from the reconstruction filter may be increased if the reconstructed sample from the encoder is lengthened before it is passed through the filter. This also has the effect of decreasing higher audio frequencies, and a post-reconstruction equaliser is needed.

Other forms of coded transmission

P.c.m. is not the only form of coded transmission. A simpler version is delta modulation. At any sampling instant the sample voltage is compared with a voltage obtained by integrating all the signals previously transmitted. If the

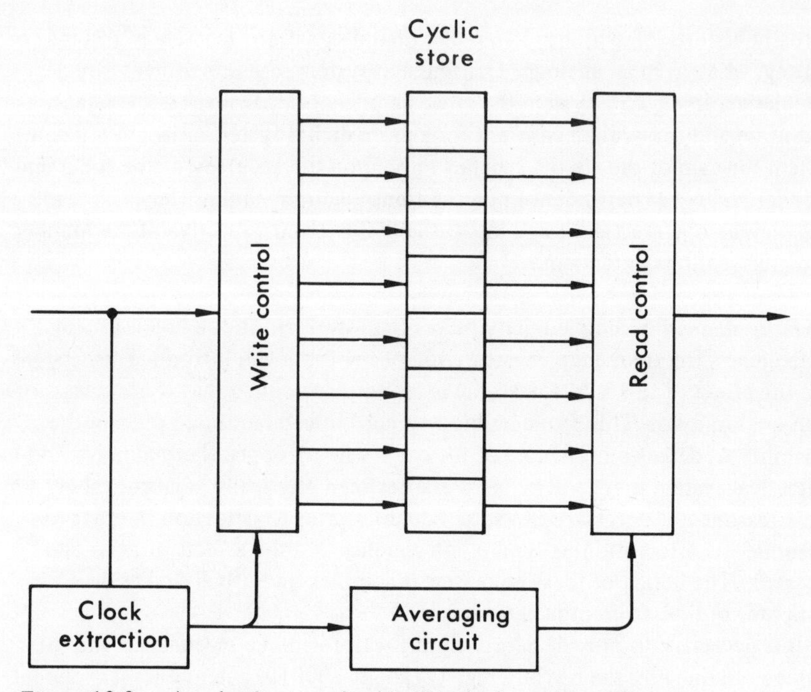

Figure 10.2 An elastic store for 'de-jittering' a pulse stream

sampled signal is greater, then a 'plus' signal is sent (e.g. '1'), if less then a 'minus' signal (e.g. '0') is sent. The demodulator need only be an integrator which is stepped up or down depending on the signal received. The basic characteristic of delta modulation is that there is a limiting rate of change of the signal beyond which the system will not follow. This rate of change will depend upon the amplitude and the frequency, and the higher the frequency, the lower the maximum amplitude of the component. It is ideally suited for speech since the majority of the energy is situated at the lower frequencies, but it would not be acceptable to signals of a more uniform spectrum such as multi-frequency signalling or data signals.

A much higher sampling rate than the Nyquist rate is necessary since the amount of information sent per sample is only 1 bit. It is found that comparable speech quality is obtained with 30 kHz sampling rate to that obtained with an 8 kHz, 7 bit, p.c.m. system, i.e. 30 kbits/s as opposed to 56 kbits/s. This is some saving in bit rate, but in view of its intrinsic inflexibility, delta modulation is unlikely to have application in a general purpose network. It may, however, have application in a cheap modulator in the local network and the code is converted to p.c.m. at the local exchange.

A compromise solution is differential p.c.m., whereby at each sample time a many-bit differential signal is sent rather than the absolute value. This can reduce the overall bit rate but it still suffers from the non-uniform spectrum-handling capacities. This is found to be useful for video work and is used on the North American picturephone system[6].

10.2 Parameters for different signals

Telephony

The parameters for a p.c.m. system for use within a large telecommunication system may be studied by consideration of carefully selected hypothetical reference connections, consisting of fairly unfavorable connections in which the losses from the lines feeding the equipment are distributed over the range of possible values. For instance, ones proposed by Richards[7], are shown in Figure 10.3; the exact parameters of a system suitable for use in an international system are still under study. Obviously any system must satisfy existing CCITT standards as regards loss, circuit noise, attenuation/frequency distortion etc. Introduction of p.c.m. systems will introduce new types of distortion, particularly quantisation noise. A sound principle for planning is therefore,

> *'Additional distortions should be so slight that they have a negligible effect upon the proportion of satisfactory calls which have previously been achieved'.*

The main parameters to be chosen are:

(a) *Sampling frequency.* The choice of sampling frequency is easy since 8 kHz allows the use of simple filters for the telephone band of 300-3400 Hz and it avoids the possibility of interference with 4 kHz channelling equipment.

(b) *Load capacity.* This is defined as the mean power level, T_{max}, of a sinusoidal signal which just occupies the range $-V$ to $+V$ defined by the coder. The required load capacity will depend upon the maximum speech volume, S dBm, at the output from a local exchange. For instance, in the United Kingdom network, average speech volume is about -15 dBm, assuming average subscriber and average line which has 7 dB send reference equivalent. For a short line the send reference equivalent is increased to 4 dB s.r.e. (i.e. 3 dB louder) so the median speech level for short lines is -12 dBm. For fixed line conditions, talkers vary in their speech volumes with a standard deviation of 2·8 dB, so it may be estimated that less than 1 percent of talkers are likely to produce speech volumes $S > 2$ dBm. The relationship between speech volume and load capacity must be found experimentally. It is found that the threshold of detectability of peak limitation corresponds to $S - T_{max} = -6$ dB (i.e. the speech volume must be 6 dB less than the load capacity). In practice the load capacity is set to a level corresponding to $S - T_{max} = -4$ dB to give a margin over this figure and also to cope with non-speech signals such as data or push-button m.f. telephones etc. This corresponds to $T_{max} = 2$ dBm0. More recent work has suggested that $T_{max} = 3$ dBm0 is preferable.

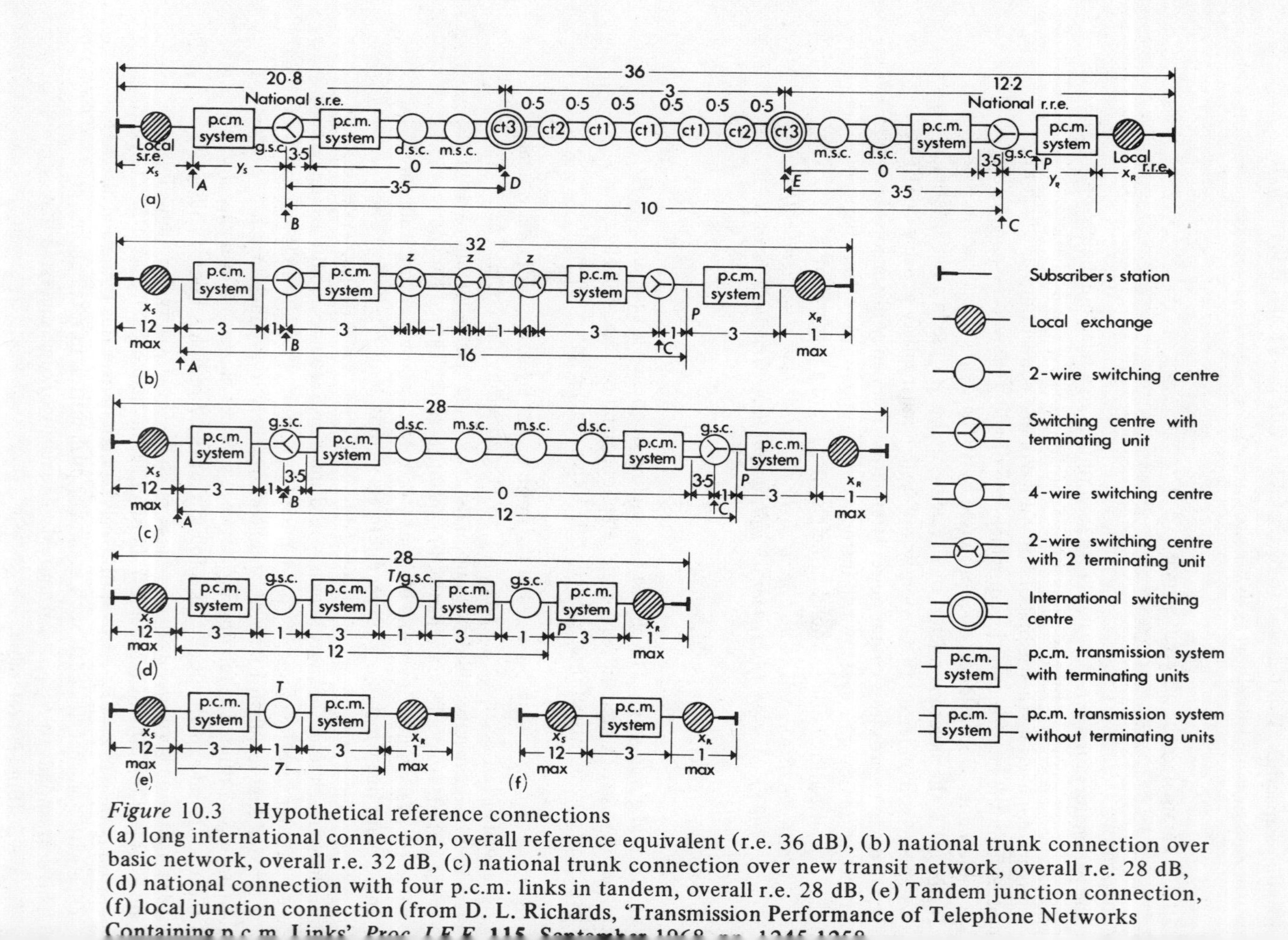

Figure 10.3 Hypothetical reference connections
(a) long international connection, overall reference equivalent (r.e. 36 dB), (b) national trunk connection over basic network, overall r.e. 32 dB, (c) national trunk connection over new transit network, overall r.e. 28 dB, (d) national connection with four p.c.m. links in tandem, overall r.e. 28 dB, (e) Tandem junction connection, (f) local junction connection (from D. L. Richards, 'Transmission Performance of Telephone Networks Containing p.c.m. Links', *Proc. I.E.E.* **115**, September 1968, pp. 1245-1258

(c) *Number of levels and companding law.* The choice of the companding law and the number of levels will effect the quantisation noise. In the middle range of speech volumes the companding will make the distortion proportional to the speech volume, and in the lower range of speech volumes the noise will be independent of speech volume. This makes the subjective assessment of this noise a complex operation. Richards has developed a technique which is based on the fact that the statistics of speech signals are approximately constant for intervals of about 20 ms (i.e. a syllabic segment). Speech may be regarded as a sequence of these segments all with different statistics. For each segment it is possible to compute the distortion power, R, which is produced and the total effective distortion will be some sort of average of the range of different R values produced by the range of syllabic segments. The way in which these distortion powers combine to produce an overall subjective effect is not known but Richards has used an empirical formulae which gives greater weight to the distortion power of the lower level syllabic segments. The resultant measure he calls Q which may therefore be regarded as the subjectively averaged value of R for a given speaker volume.

With the aid of this sort of measure it is possible to compare the effects of different companding laws. There are now two in common use for telephone systems. They may be described as a relationship between the instantaneous input voltage and the quantisation level.

If v = the instantaneous input voltage,
V = the maximum input voltage for which peak limiting is absent,
i = the number of the quantisation level (numbered from the centre of the range)
and B = number of quantisation levels each side of the centre,
then the two laws may be expressed:

(i) *A law*

$$\frac{i}{B} = \begin{cases} \dfrac{Av/V}{1 + \log A} & \text{for } |v| \leqslant \dfrac{V}{A} \\[2ex] \dfrac{1 + \log Av/V}{1 + \log A} & \text{for } \dfrac{V}{A} \leqslant |v| \leqslant V \end{cases}$$

(ii) *μ law*

$$\frac{i}{B} = \frac{\log(1 + \mu v/V)}{\log(1 + \mu)}$$

Both these laws have the property that they are linear for small values of v/V and logarithmic for higher values [8]. In the United Kingdom a 7-bit system with a companding law of the first type is used with $A = 87{\cdot}6$. This gives a companding advantage of 24·1 dB. In North America the μ-law is used with $\mu = 100$ which gives a companding advantage of 26·7 dB.

In practice most companding is produced by a piece-wise linear approximation to the theoretical companding law and this has the effect of slightly increasing the quantising noise.

More recently, 8-bit systems are being designed to provide a higher quality link which may be used for trunk and international working.

(d) *Idle channel noise.* P.c.m. systems will generate noise in the absence of signals if the lowest decision levels in the encoder drift with time. Also an amplification effect on noise is produced if a noise voltage is present when the input level to the encoder is near a decision level. In this case a small change on input voltage will produce noise voltages of one quantum level. These factors give constraints on the stability bounds of the lowest decision levels of the system.

Programme or music channels

For broadcast quality transmission it is necessary to transmit frequencies up to 15 kHz and to achieve a peak-to-peak signal to r.m.s. noise ratio of 73-78 dB. This corresponds to a linear encoding of 11 or 12 digits (which gives a theoretical 77 dB or 83 dB quantisation noise). It is not possible to use any pre-emphasis on a music circuit since a cymbal can, for instance, produce a high-amplitude high-frequency component whose limiting would be objectionable. Companding is not used as it would not give a significant advantage. It would anyway be difficult to obtain an adequate matching of the non-linear circuits necessary. The alternative technique of using more digits and digital operation is also impractical since it is difficult enough to obtain even 11 digits, let alone 12. At present this type of application is still in the development stage, but it will be needed in the future as the ready availability of good copper circuits is reduced due to the increased use of microwave and p.c.m. systems.

Another application is the use of p.c.m. to send a television sound channel within the 4 μs blanking period of a synchronous signal. This obviates the necessity of a completely separate sound channel as is used on a programme distribution circuit, and this technique is being introduced into the United Kingdom television distribution network.

F.d.m. signals

It is possible that direct p.c.m. coding of a f.d.m. signal might be necessary for the same reasons as advanced for television signals. Bell Laboratories have experimented with direct coding of a master group (600 channels–2·4 MHz bandwidth). Since the signal is already bandlimited then a 6 MHz sampling rate was found adequate and 9-bits linear encoding was found necessary to achieve an adequate intermodulation performance when there are likely to be several stages of coding and decoding.

Television signals
There are applications for which the digital transmission of television signals is advantageous. These are for long-distance transmission, or for transmission over an essentially digital medium such as glass fibres or laser beams, or within a p.c.m. multiplex system. Also, once signals are in digital form it is relatively simple to perform operations such as mixing, aperture corrections and line conversion. For the future, use of digital techniques would considerably simplify the present procedure for setting up video links.

A 625-line, 50 Hz colour television signal has a bandwidth of 5·5 MHz and a sampling rate of about 13 MHz is necessary in order to allow for the finite rate of cut-off of a practical filter with acceptable delay characteristics. It is found that adequate visual quality is obtained with 7 or 8 bits but more would be needed if there were the possibility of more than one stage of coding and decoding since the quantisation noise would be additive (on a power basis). This implies a digit rate of around 100 Mbits/s and very fast digital circuitry.

Picturephone system
The picturephone signal used in the Bell system has a 1 MHz bandwidth with a line frequency rate of exactly 8 kHz. The sampling rate is just over 2 MHz and is synchronised to this line rate in order to sample the picture with a uniform grid. Since in a video signal there is some correlation between different samples, a differential p.c.m. system of coding may be used, and from subjective tests it is found that a 3-bit code gives adequate results, i.e. a 6 Mbits/s rate.

10.3 Applications of p.c.m.

Junction network
The major application for p.c.m. techniques is in the junction network using existing audio cables. The attenuation of these cables increases with frequency and this limits their usefulness for f.d.m. to around 100 kHz. However, because a digital system using frequent regenerators can work at a very poor signal-to-noise ratio, it is possible to use an audio pair up to a very high digit rate. A convenient location for regenerators is in place of loading coils which are normally placed (where used) every 1600 metres. At this distance it is found that digit rates of up to 2 Mbits/s are practical, the main limitation being cross-talk from other digital systems in the same cable. The most suitable system to this digit rate was found to be a 24-channel multiplex system with a 7 bit code and an additional bit for signalling purposes. With an 8 kHz sampling rate this yields a total bit rate of 1·536 Mb/s. Hence, two 2-wire audio circuits can be converted to 24 4-wire circuits, i.e. a gain of 22 circuits. In Europe there are currently discussions aimed towards a 32-channel system working at 2·048 Mbits/s. This system would use 30 channels for 8-bit speech samples and the remaining 2 channels for signalling and synchronisation information.

The cost of a p.c.m. system is primarily in the terminal equipment, and at the present time 24 channel p.c.m. systems are cheaper than audio pairs for junction routes in excess of about 16 km. This figure is coming down as the cost of the terminal equipment is reducing.

Framing

At the demultiplexer it is necessary to be able to sort out which bit is which, i.e. establish a frame. For the 24-channel system a frame consists of 24 x 8 = 192 bits. In the Bell system framing information is obtained by adding an extra bit per frame and producing a unique pattern with this bit. At the demultiplexer the incoming bit stream is searched for this pattern which, once found gives the frame information. If the pattern disappears then a reframing operation must be initiated. Obviously the probability of reframing and the time taken to re-establish the frame must both be low since communications will be interrupted whilst it occurs.

The British system has a different philosophy for framing and uses the signalling bits to send the information. In this system a multi-frame format is used consisting of 4 frames. Every first frame of the 24 signalling bits (one per channel) is used for a unique framing pattern. Every second and fourth frame, the signalling bits are used as normal and the third frame is redundant. The advantage of this latter system is that there are 24 bits of framing information every 4 frames, whereas in the Bell system there is only 1 bit per frame. Hence, reframing will take a shorter time with the British system, but a cost of more complicated circuitry and a reduction in the flexibility of the system.

Digital switching

A significant reduction in economic distance will occur if the signal is switched in its digital form at the tandem or trunk exchange [9], since this will save a decoder and coder. An improvement in the transmission will also result, since the intermediate exchange (or exchanges) will not introduce any additional distortions, and a complete digital link may be operated at a fixed loss around 3 dB however many digital switches there may be.

The major problem of digital switching, though, is that it usually implies that the system must be synchronised as asynchronous p.c.m. switches require a large amount of storage and are relatively expensive. The problem of synchronising, in a reliable manner, large parts of a country has not yet been completely solved satisfactorily.

10.4 Higher multiplex systems

It is necessary to produce a hierarchy of multiplex systems for p.c.m. in order to use high-speed digital links [10]. In f.d.m. systems the identity of each channel is easily maintained in the multiplex signal because of its unique frequency allocation. In a time division multiplex system there are the problems of synchronisation and framing to identify which bit goes where at the end of the demultiplex part of the system. Framing is achieved in similar manner to the technique used in the 24-channel system.

The main problem is that of combining many bit streams arising from independent sources. It is neither practical nor desirable to insist on synchronising all the sources to some universal rate, and to allow asynchronous working a technique known as *bit stuffing* is employed [11]. With this technique the frame rate is chosen to be slightly higher than the maximum rate of the channels to be multiplexed. Hence, every so often there will be no bits to be transmitted and a zero pulse train is substituted. Part of the frame is used as a control channel, and at the time of inserting the zero pulse train a code is sent on the control channel to indicate that stuffing has occurred and on which channel. At the demultiplexer end it is then possible to ignore the stuffed pulse. This means that the outgoing pulses from a demultiplexer will occur in short regular sequences with gaps in between, and it will be necessary to put them through an elastic store to smooth the rate out.

The Bell system proposals (Figure 10.4) for their multiplex system takes the basic building block to be their 1·544 Mbits/s 24 channel system (T1 system), and the first stage is to take four of these to make a 6·312 Mbits/s signal (i.e. slightly higher than 4 x 1·544 Mbit to allow for framing and stuffing). This rate (called a T2 system) is also suitable for direct encoding of the picturephone signal. The next stage in the hierarchy is T3 signals consisting of 46·3 Mbits/s which corresponds to seven T2 signals which is also suitable for the direct encoding of 600 channel f.d.m. signals. Broadcast television signals need 92·6 Mbits/s. Groups of these 46·3 Mbits/s signals are then combined to form rates of several hundred megabits for transmission to line.

Similar systems are under consideration in Europe and will be based on the 32-channel system (2·048 Mbits/s) as the basic building block.

10.5 Application to other media

Although the original p.c.m. equipment was designed for use on twisted pair cables, it is possible to use it for many other media, especially if high speeds are needed.

(a) *Coaxial cable* A coaxial cable has a very good signal to noise ratio and a high attenuation to crosstalk path. It is therefore an ideal media for high speed pulse transmission and systems working up to 1 Gbits/s (i.e. 1000 Mbits/s) have been constructed. With these systems it is possible to use a number of repeaters between the regenerators, and it turns out that the optimum distance for repeaters is the same as if the cable were being used for f.d.m. Hence, it is possible to convert a cable system to pulse use by replacing a proportion of the repeaters by regenerators. There may also be an advantage in using multi-level coding to increase the bit rate.

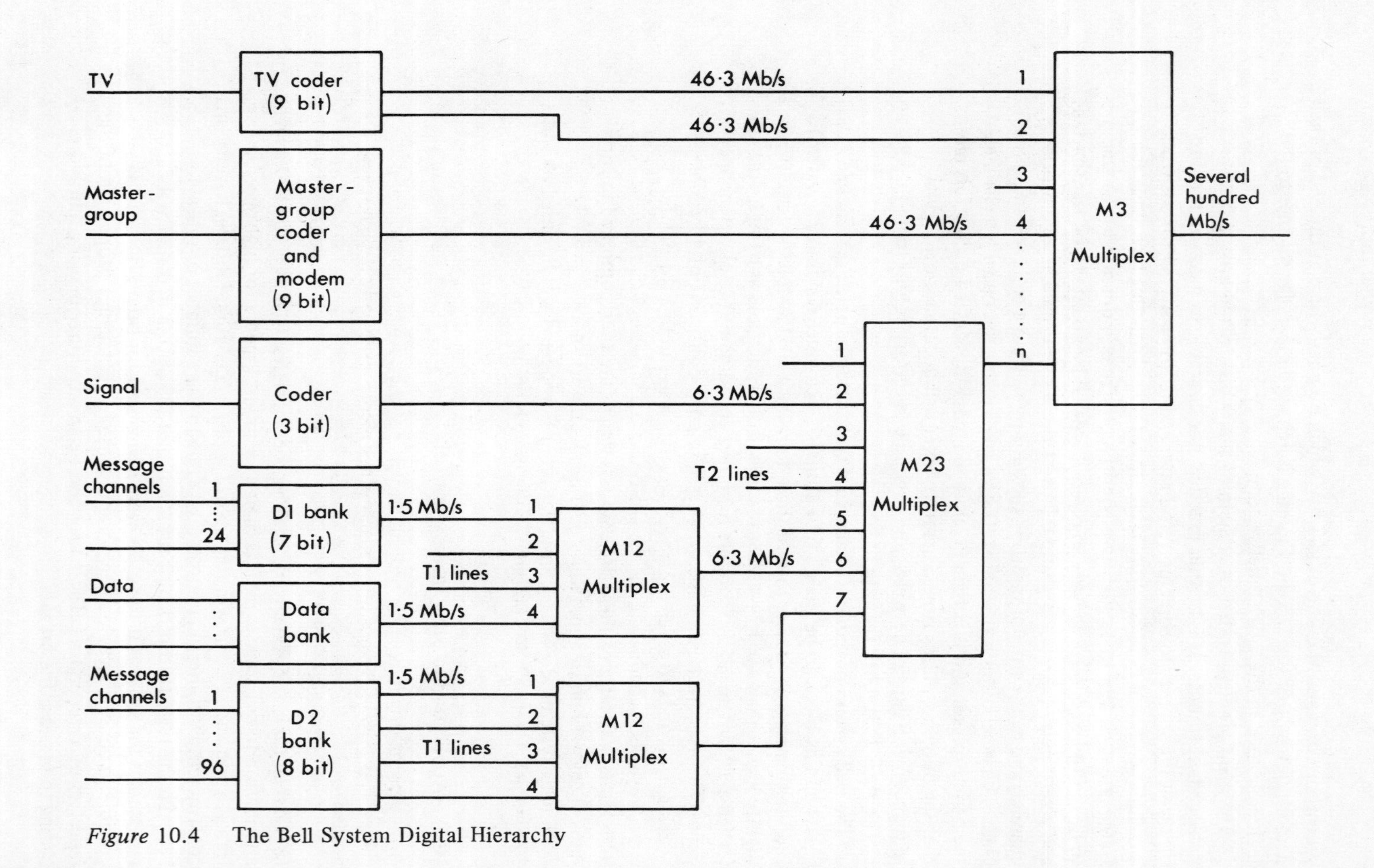

Figure 10.4 The Bell System Digital Hierarchy

(b) *Microwave systems* A given signal in p.c.m. form will occupy a greater bandwidth than in f.d.m. form and therefore it might be thought that the use of p.c.m. for a radio system would be uneconomic. This is not, however, the case, since if a line-of-sight microwave system is used digitally, then the same frequency allocation may be used for each leg. Thus an overall saving in frequency space is made and the intermediate repeaters are more efficient and of simpler design. The cost of the terminal equipment is, however, high.

(c) *Satellite systems* P.c.m. is also advantageous for satellite systems, especially when they are to be used for multi-access, since a t.d.m. system implies that the full carrier power of the satellite is available for every signal (see Chapter 8).

REFERENCES

1. For example, see G. C. Hartley, *Techniques for Pulse Code Modulation*, CUP, 1967.
   or, for a more detailed description of the basic theory, see K. W. Cattermole, *Principles of Pulse Code Modulation*, Iliffe, 1969.
2. B. Smith, 'Instantaneous Companding of Quantised Signals', *B.S.T.J.* **36**, May 1957, pp. 653-709.
   See also
   K. W. Cattermole, *Principles of Pulse Code Modulation*, section 3.2, Iliffe, 1969.
3. D. L. Richards, 'Transmission Performance of Telephone Networks Containing P.C.M. Links', *Proc. I.E.E.* **115**, 9, September 1968, pp. 1245-58.
4. For detailed discussion see Transmission Systems for Communications, Western Electric, 1970, Chapter 27.
5. M. R. Aaron and J. R. Gray, 'Probability Distributions for the Phase Jitter in Self-Timed Reconstructive Repeaters for P.C.M.', *B.S.T.J.* **41**, March 1962, pp. 503-58.
6. H. I. Mansell, and J. B. Millard, 'Digital Encoding of the Video Signal', *B.S.T.J.* **50**, Picturephone issue, 2, February 1971, pp. 459-480.
7. D. L. Richards, 'Transmission Performance of Telephone Networks Containing P.C.M. Links', *Proc. I.E.E.* **115**, 9, September 1968, pp. 1245-58.
8. For a detailed discussion of their relative merits see K. W. Cattermole, *op. cit.*, section 3.2.
9. K. J. Chapman and C. J. Hughes, 'A Field Trial of an Experimental Pulse Code Modulation Tandem Exchange', *Post Office Electrical Engineers Journal* **61**, October 1968, pp. 186-95.
10. J. S. Mayo, 'Experimental 224 Mb/s P.C.M. Terminal', *B.S.T.J.* **44**, 1965, pp. 1813-41.
11. F. J. Witt, "An Experimental 224 Mb/s Digital Multiplexor using Pulse Stuffing Synchronisation', *B.S.T.J.* **44**, 1965, pp. 1845-86.

Chapter 11

Telegraphy systems

11.1 Introduction

Telegraphy was the original technique of electrical communication and used simple Morse-code sounders and relays over a single wire with an earth return. The first stage in automation was the use of a Morse printing machine which put the dots and dashes on to a strip of paper. The use of Morse-code obviously requires skilled operators and the first successful letter printing machine was invented by R. E. House in 1846; this formed the basis of the 'ticker-tape' machine[1].

In this machine there is a wheel with embossed letters around the periphery. Normally the wheel is at rest, but on receipt of a positive current the wheel is released and it rotates at a constant velocity until a negative current pulse arrives. At this time the wheel drops and prints the particular letter at the bottom of the wheel. After printing the wheel continues rotating until it stops at its rest position. Thus the letter that is printed is controlled by the time between the arrival of the two pulses. The system worked at around 70 characters/minute.

This is an absurdly simple system and is very small, cheap and quiet. The power can even come from a clock-type weight. The generator normally takes the form of piano-type keyboard with a key per letter. The main limitation is that of speed of transmission, since the wheel must make a complete revolution for each character. These devices are still in use today for a cheap, quiet, information dissemination service. For instance, until recently they were used at London Airport to disseminate details of aircraft arrivals and departures to the airline offices and services.

The slow transmission speed can be increased if the required characters are sent in some form of code. The one used nowadays is based on a 5-unit binary code giving a total of 32 symbols. This is insufficient for the 26 letters of the alphabet, numerous punctuation and control characters, so a 'letter-shift' and 'figure-shift' are used to increase the capacity up to 62 characters. The use of such a code means that the transmitter and receiver must be in some form of synchronism so that the receiver can identify each code unit. The simple technique used for achieving this is by the use of what is called start/stop signals. When a teleprinter is operational it can send either of two conditions, a 'mark' or a 'space'. In the United Kingdom a mark is sent as +80 V and a space by –80 V, and they are detected by means of a polarised relay or suitable transistor circuit. (When used within a switching system such as Telex, a 0 V signal is also used for supervisory purposes and this introduces coding problems as will be seen later). This is an example of what is called *double-current working*, and allows a faster transmission rate and more reliable detection than *single-current working* using only the presence or absence of a potential.

In the United Kingdom system [2], when a teleprinter is switched on but is not actually sending a character, a continuous 'mark' is transmitted. Before any character is sent, a 20 ms 'space' is transmitted, indicating the start of a character. There are then five 20 ms time-slots in which a 'mark' or a 'space' may be transmitted and these give the five code digits (see Figure 11.1). These code digits are then followed by a 'stop' signal of at least 30 ms consisting of a 'mark'. Hence the transmitter and receiver (which in general are mechanical devices) need only be in synchronism for the duration of one character. The necessity for a 'stop' signal of a different duration to that of the other units is to allow it to be identified in the case of continuous transmission, as might occur if pre-punched paper tape were used rather than a human typist.

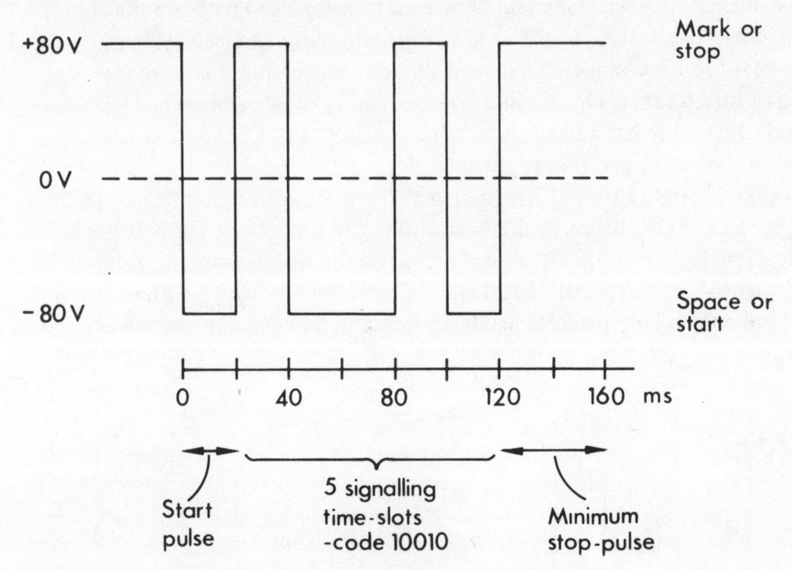

Figure 11.1 Start/stop format for the 5-unit code

With this system, the minimum character period is 150 ms and the start/stop code is sometimes referred to as a 7½ unit code. The maximum rate of information transfer comes to 400 characters/minute. Since the minimum bit width is 20 ms, then the channel capacity needed is 50 bauds. These parameters are standardised throughout Europe but in the United States of America the bit element is 22 ms, giving rise to a 45·5 baud system. The stop signal duration is 1·4 units so that the total character cycle is 163·3 ms. This obviously introduces compatibility problems when interconnection is required between the two systems.

These speeds were governed by the mechanised tolerance obtainable on the early teleprinters, together with estimates of the fastest typing rates anticipated. Nowadays higher speeds, up to 100 baud working, are used for operation with paper-tape.

Because of the different ways in which mark/space, start/stop, ±80 V etc. are used in different countries, the CCITT have standardised the use of the symbol A for a 'start' signal and Z for a 'stop' signal.

Prior to 1929, the teleprinters used printed tape, but in 1929 the now familiar page teleprinter was introduced and the CCIT (as it then was) standardised the alphabet used. This standardisation, so early in the history of teleprinters, has meant that it is possible to intercommunicate between machines almost anywhere, without any code changing being necessary.

Transmission systems

Teleprinter systems may be operated on a *simplex* (one direction of transmission at any one time) or a *duplex* (simultaneous transmission) basis. In the simplex arrangement it is satisfactory to use a single wire with an earth system, provided a low-pass filter is connected between the transmitter and line to reduce the effect of interference. One duplex arrangement is shown in Figure 11.2 where the polarised relay has a balance winding connected to a balance impedance so that simultaneous operation is possible.

In practice in the United Kingdom a two-line simplex system is used with a local power supply, although this is equivalent to a duplex system, it is not usable as such since normally a local record of the transmission is required.

In the early days (pre-1930) all circuits were on a private wire basis using mainly phantom circuits. This had a considerable effect upon loaded sections

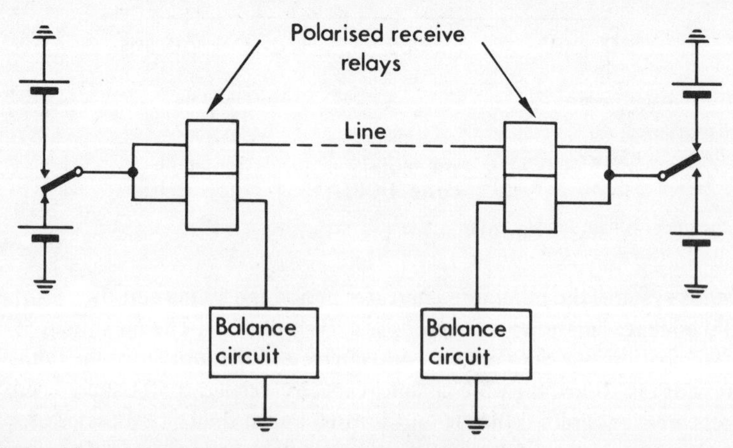

Figure 11.2 Duplex telegraph transmission

since it meant that the loading coils must not change their characteristics when a signalling current passed through them. When the circuits had to use a carrier system, the signals were sent by the on/off keying of a 300 Hz tone (i.e. 100 per cent amplitude modulation).

In the 1930s, multi-channel voice frequency (m.c.v.f.) systems were introduced, which give up to twelve telegraph channels within a single audio channel. In 1932 the British Post Office took the decision to introduce a switched teleprinter service called Telex (*Tel*egraph *Ex*change). This was to be based on the existing telephone network and used v.f. senders. The war halted development of this system, and afterwards, in view of the different traffic properties and requirements of the Telex system, it was decided to develop on a separate network.

Distortion of Pulse

The effect of a finite response time of the transmission network plus adjustment of the polarised relay etc. means that the pulse obtained from the output contacts of the receive relay will be distorted and there will be a difference between the element duration as transmitted and received. The percentage difference is normally referred to as *telegraph distortion*. This can vary from instant to instant and also with the actual pattern of transmitted pulses, and the figure quoted is normally that for the worst case. The distortion is normally classified into three types:

(a) *Characteristic distortion* – produced by the transient response of the transmission media.
(b) *Fortuitous distortion* – caused by irregularities in any part of the circuit or by interference.
(c) *Bias distortion* – a consistent lengthening of one of the elements due to some asymmetry in the transmitting or receiving equipment.

A transmitter is normally adjusted so that it has less than 5 per cent distortion and distortions up to 40 per cent may still be recoverable by a well adjusted receiver. It is possible to use regenerative repeaters in order to correct bad degrees of distortion and the CCITT recommend a maximum degree of distortion for any complete circuit to be 28 per cent, whether they are equipped with regenerative repeaters or not [3]. Individual design values are also given by the CCITT.

Transmission planning for a telegraph circuit consists of allocating the total distortions to the constituent parts of the overall connection [4]. It is found that the characteristic distortions of individual links add together linearly for a multi-link connection but that the bias and fortuitous distortions add on a r.m.s. basis. Hence the overall distortion of a connection is given approximately by

$$\delta_{tot} = \sum_{i=1}^{n} \delta_{char} + \left[\sum_{i=1}^{n} (\delta_{bias})^2 + \sum_{i=1}^{n} (\delta_{fort})^2 \right]^{1/2}$$

11.2 Multiplexing techniques

Frequency-division amplitude-modulation

One of the earliest multiplex systems was introduced in 1925 and used six keyed voice-frequency (v.f.) tones, and separated them at the remote end using constant-k filters. With improved filter design and better frequency response of the trunk network, the number of telegraph channels in one audio channel was increased to 12 and then 18 channels, using 120 Hz spacing. Some 24-channel equipment is also in use today.

An on-off v.f. signal is equivalent to a 100 per cent amplitude-modulated signal modulated with a square wave, and since the spectrum of a square wave has a large number of harmonics, then a very wide spectrum of side-bands would be produced. In order to reduce the number of sidebands, the keyed v.f. signal is passed through a suitable band-pass filter before being added to its fellow travellers on the multiplexed path. At the far end of the system the signals are separated by further band-pass filters.

The effect of passing a keyed v.f. signal through a band-pass filter, thereby curtailing the side-bands, is to delay the build-up of the full carrier amplitude. It can be shown [5] that if t is the build-up time in seconds for a carrier at the centre frequency of the filter, and f_1 and f_2 are the effective cut-off frequencies in Hertz, then

$$t \simeq \frac{1}{f_2 - f_1} \text{ s}$$

Hence a signal must have a duration of at least t seconds to be correctly received. The limiting speed on bauds given by

$$N = \frac{1}{t} = f_2 - f_1 \text{ bauds}$$

This is, of course, only an approximate result, and in practice somewhat wider bandwidths must be used in order to cater for variations in carrier frequencies and the use of non-ideal band-pass filters. For instance, the standard CCITT system recommends the use of 120 Hz spacing for 50 baud signals. In theory, the signalling rate is limited only by the noise, as shown by Shannon's equation, and in some systems the use of a more sophisticated modulation and detection system allows a much higher bit rate than this simple approach indicates (see Chapter 12).

Choice of carrier frequencies

The effect of any non-linearity in the amplitude characteristics of the transmission path will introduce intermodulation products and spurious harmonics. (Remember that on an individual speech channel, a considerable amount of non-linearity is permissible before it becomes noticeable on the speech.) If f_1 and f_2 are any two carrier frequencies, then second-order intermodulation products will occur at $2f_1$, $2f_2$, $f_1 \pm f_2$ and third-order products at $2f_1 - f_2$, $3f_1$ etc. In general it will be the second-order products that will have the largest amplitude, so if all the carrier frequencies are chosen to be *odd* multiples of some common frequency, then all the second-order intermodulation products will be of *even* order of this common frequency. The products thus fall between the carrier frequencies and the band-pass filters may be arranged to attenuate them. In all the CCITT recommended systems the common frequency is chosen to be 60 Hz, giving 120 Hz between the carriers.

In early equipment, the v.f. tones were all produced by an electric motor rotating at 60 Hz and switched with a relay or else with a diode switch, but modern equipment produces the tone electronically.

Frequency-division frequency-modulation

With the achievement of transistorised circuits it has been possible to use frequency modulation rather than amplitude modulation [6]. This has the usual advantages of f.m. over a.m. of

(a) operates over a wide range of received signal level,
(b) improved performance with given signal-to-noise ratio,
(c) rapid response to changes in signal level,
(d) higher speed of operations for a given channel spacing.

This type of system is of special usefulness where transmission channels are noisy or where transmitted power must be limited. These systems are now used almost universally for new installations, and use the same carrier frequencies as the a.m. system with ±30 Hz frequency shift [7].

Time-division-multiplex

The main advantages of t.d.m. are

(a) peak power of carrier always available,
(b) possibly simpler circuits than a.m. or f.m.,
(c) sub-multiplexing possible, i.e. ½ or ¼ channels can be produced;

and the disadvantages are:

(a) requires storage and some synchronisation,
(b) multipath propagations in radio and phase distortion in, cables are limiting factors.

For instance, in radio systems, there can be path differences of up to 4 ms, so that a minimum element must be about 10 ms, i.e. speed is limited to 100 bauds. The phase distortion of an audio channel in a submarine cable limits the speed of a 120 Hz spacer carrier to about 80 bauds.

Hence, assuming that terminal cost is not the limiting factor, the most efficient solution is to use as much time-division as possible and then use frequency-division for the rest of the multiplexing.

The use of t.d.m. implies synchronous working and in this case there is no need for the 'start' and 'stop' signals, so that a simple 5-unit code could be transmitted. However, when teleprinters are used in a switched network, such as Telex, then a continuous-start or a continuous-stop have supervisory functions, so that if these were transmitted and at the remote end of the t.d.m. part of the system the 'start' and 'stop' pulses were re-inserted, then the supervisory signals would then appear to be conventional characters. For this reason, a 6-unit code is needed.

In order not to miss any character, it is necessary to make the basic speed of any t.d.m. system higher than that of the highest speed teleprinter that is likely to be connected to the system. The CCITT have recommended a character length of $145\frac{5}{8}$ ms for use in t.d.m. systems which corresponds to $411\frac{2}{7}$ characters/minute, and hence such systems will accept 50 baud (400 characters/ minute) nominal signals up to 3 per cent fast. If two 6-unit code channels are then operated at $411\frac{2}{7}$ characters/minute, the aggregate speed is 82 bauds.

The channels can be interleaved on a character or an element basis; the latter requires less storage of the signals prior to transmission. The element synchronisation of the two ends may be achieved by the transitions of the received elements themselves, and frame synchronisation is obtained by means of special codes sent with the aid of the sixth bit [8].

This type of equipment may be installed using any v.f. channel capable of working at 82 bauds without any change in the channel power loading. In particular it is employed to increase the number of telegraph channels in the Transatlantic cables.

In a submarine cable the speed limitation is produced by the restricted bandwidth per channel and it is possible to use automatic compensation at the receiving end to reduce this effect and increase the speed to nearer the theoretical maximum of twice the bandwidth. The equipment to do this is called a *characteristic distortion compensator* (c.d.c.) and permits a conventional f.m.v.f. channel to take 3 t.d.m. channels [9]. The principle of operation of this device may be understood by considering first the manner in which a conventional synchronous systems works. If there is no bandwidth limitation then the detector will sample the received signal and give a mark or space output depending whether the signal is above or below a zero voltage decision level (as shown in Figure 11.3a).

On a channel of finite bandwidth, as the modulation rate is increased the build-up time of the signal change becomes more significant until a point is reached where the signal may no longer be reliably detected (see Figure 11.3b). The c.d.c. equipment overcomes this problem by moving the decision level depending upon the last detected signal as shown in Figure 11.3c.

Modulation systems for h.f. radio

In an h.f. radio system it is usually impractical to employ much time-division, and the use of frequency division divides the available carrier power amongst

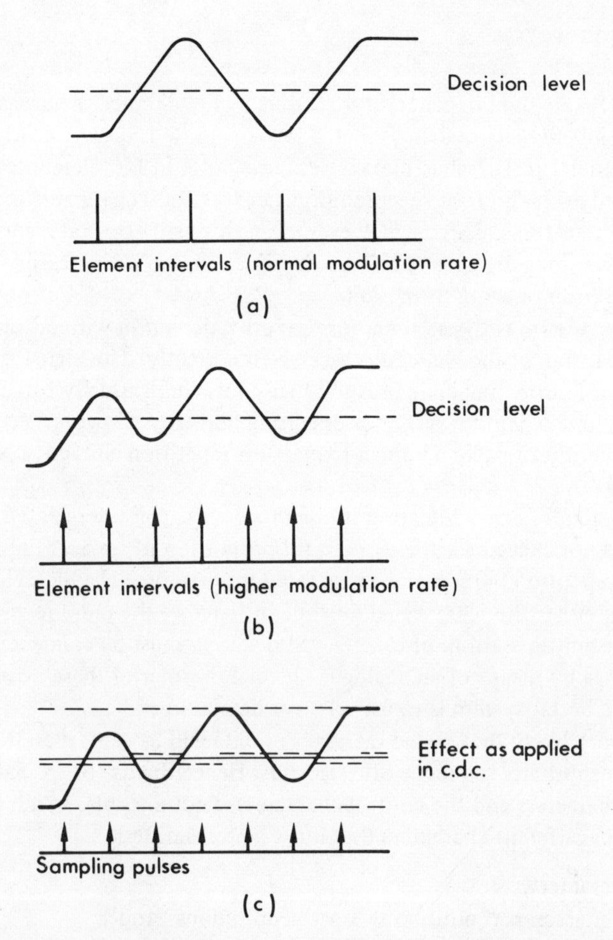

Figure 11.3 Principle of the characteristic distortion compensator
(a) simple detection, (b) the effect of bandwidth limitation, (c) movement of decision level to improve detection

the sub-carriers. The channel capacity of an h.f. link may be increased by use of what is called *diplex* modulation. In this system the state of two channels is coded on to one out of four separate carriers. This ensures that full power is always available at the expense of using more bandwidth. Note that it is not necessary for the two channels to be synchronised.

Another technique for use on h.f. systems is the 'Piccolo' system[10]. This is used for very noisy circuits and works by taking the 5-unit code and converting it to one out of thirty two carriers. These carrier signals are now present for 7½ times as long as a single element and are detected by a set of matched filters at the receiver. This is another example of the trade-off of increased bandwidth to produce a better signal-to-noise behaviour.

11.3 Error control

Since teleprinters are used in general for sending highly detailed messages without much inherent redundancy, then a low error rate is desirable. Ideally a figure of the order 1 in 10^5 characters is desirable but 1 in 10^4 elements is usually regarded as the limit of acceptability (i.e. 1 in 2000 characters for a 5-unit synchronous transmission). However, even this high error rate is usually unachievable over long distance radio circuits. The rate can be increased by means of some form of error correction.

The simplest system is to use some form of error *detection* with automatic request for repetition of the characters received incorrectly. This signal is called ARQ. A full error-correcting code uses too many bits and is usually too expensive in channel time when the system is operating normally. However, errors occur too frequently to make a human request for repetition a practical solution.

A simple system of error detection is one in which signals are rejected if their amplitude does not exceed a set level for a set proportion of time. A failure here could cause a repetition request to be generated in the return channel. However, this means that either the ARQ signal must be initiated at the radio receiver and sent to the transmitter station, or else the radio station must be connected to the control exchange by means of an analogue channel. Neither of these possibilities are attractive; a better system is to use an error-detection code.

Since it is very likely that coded messages or data will be sent, then it must be possible to transport any sequence of characters. Hence the use of special sequences of characters and the control characters is not possible. In all, there are a total of 35 different characters that must be transmitted:

32 basic characters,
2 control characters (continuous 'start', continuous 'stop'),
1 request for repetition (ARQ signal).

For obvious reasons, there must be simple conversion between the required code and its error detection form and the number of extra elements must be as small as possible so as to minimise transmission time. Also the errors in the same sense should not cause undetected errors. Although a minimum of 6 elements is sufficient for the 35 code combinations, it is not possible to provide any error detection. Hence a 7-unit code must be used and a constant weight code of 3 marks and 4 spaces provides 35 combinations and allows simple error detection which can now be performed at the control exchange and digital link may be used between the receiver and the exchange.

One such system, invented by Van Duuren, has been standardised by the CCITT in 1956[11]. This is normally applied to 2- or 4- channel t.d.m. character interleaved systems with 'go' and 'return' paths. At the control exchange, incoming signals are stored on magnetic drums. Each character is then transmitted, and if a mutilation is detected at the far end then the printing is stopped for four characters and an ARQ signal is sent on the return channel as soon as timing permits, and is followed by the last three characters previously transmitted (see Figure 11.4).

At the receiver, when the ARQ signal is detected, it suspends normal transmission and sends instead an ARQ signal followed by the last three characters that it transmitted. Since the reception of the original incorrect character had

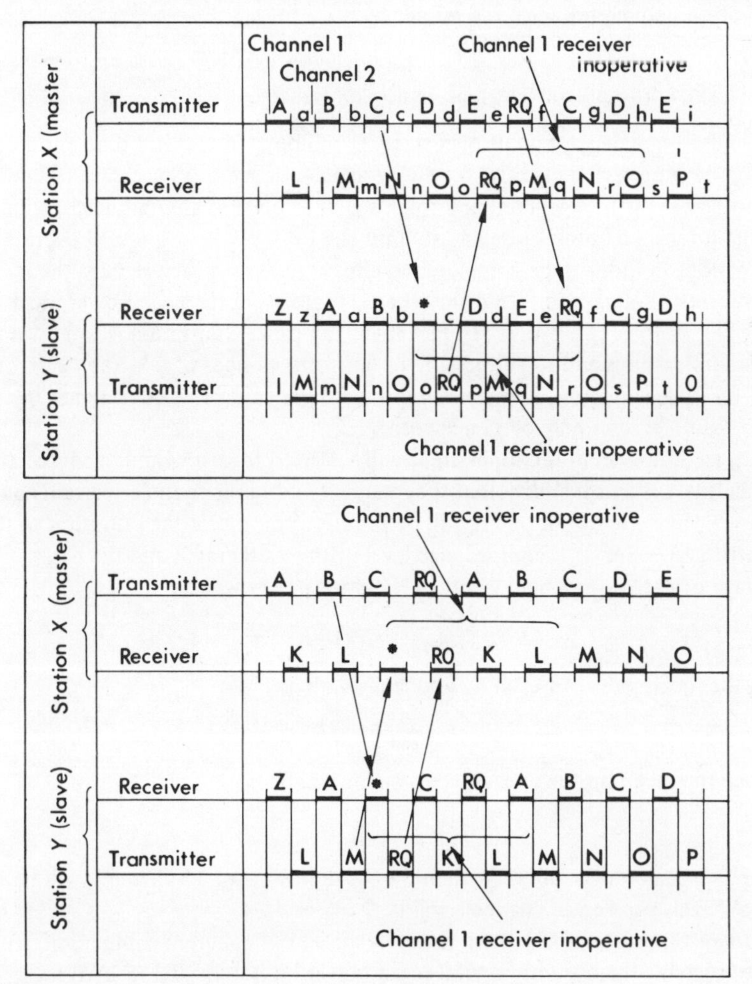

Figure 11.4 Operation of the ARQ system
(a) simple mutilation in one direction of transmission (b) simultaneous mutilation in both directions of transmission. The master station controls the telegraph speed of the whole system. The slave station must remain in a locked time relationship with the master station both when receiving and transmitting (taken from A. C. Croisdale, 'Teleprinting Over Long Distance Radio Links–II', *Post Office Electrical Engineers Journal* **51**, 1958, pp. 219-225.

suspended printing at the remote end for four characters, then the first character to be printed again (assuming no subsequent error in either channel) will be a copy of the mutilated character.

A careful study of Figure 11.4 will show that the system will still work satisfactorily in the event of repeated errors or errors in the ARQ signal sent on the return channel. A detailed description of the complete apparatus for this operation may be found in the references.

11.4 Power levels for telegraph multiplex signals

In general the telegraph multiplex systems will be used within f.d.m. systems which have been designed to take audio signals. Hence it is necessary to limit the mean power of the telegraph multiplex signal to prevent any undue contribution to the intermodulation noise and to limit the peak power to prevent overload distortion. In order to be completely compatible with speech signals, the mean power must not exceed –15 dBm0 (i.e. 31·6 μW) and the peak power must not exceed +7 dBm0 (i.e. 5 mW). In fact the mean power limitation is too severe for most applications and it is necessary to limit the number of telegraph systems in any f.d.m. group (typically not more than 20 per cent of voice channels in a large f.d.m. system may be non-speech)[12].

In an a.m. system the tones are usually derived from a common source and hence there is a probability that they may all add in phase under certain conditions. Hence the peak power will be produced by the voltage addition of the constituent tones. If the impedance level of the system is R and the r.m.s. voltage of each component is v then the peak power will be

$$p_{\text{peak}} = (Nv)^2/R$$

and the mean power of each component will be

$$p_{\text{ind}} = v^2/R$$

Hence the mean power of each component will be

$$p_{\text{ind}} = p_{\text{peak}}/N^2$$

For a peak power of 5 mW then the mean power for 12 channels will be 35 μW each and for 24 channels will be 9 μW each.

In an f.m. system[13] there is less likelihood of coherent summation, and in these systems there is only a total mean power limitation of 135 μW0 for however many channels there are, with the limitation that for 12 channels or less the mean power does not exceed 11·25 μW0 each.

REFERENCES

1. For example see the history in 'Telegraphy', *Post Office Electrical Engineers Journal* **49**, 1956, pp. 166-72.
2. J. W. Freebody, *Telegraphy*, Pitman, 1958.
3. *C.C.I.T.T. White Book VII*, Recommendation R50.
4. R. W. Barton, *Telex*, Pitman, 1968, Chapter 4.
5. For example see J. Brown and E. V. D. Glazier, *Telecommunications*, Chapman ahd Hall, 1964, pp. 128-32.
6. W. F. S. Chittleburgh, 'A Frequency-Modulated Voice-Frequency Telegraph System', *Post Office Electrical Engineers Journal* **50**, 1957, p. 69.
7. *C.C.I.T.T. White Book VII*, Recommendation R35.
8. Details may be found in *C.C.I.T.T. White Book VII*, Recommendation R44.
9. A. C. Croisdale and C. S. Hunt, 'Synchronous Multiplex Telegraphy on Intercontinental Submarine Telephone Calls', *Post Office Electrical Engineers Journal* **60**, 1, 1967, p. 52.
10. H. F. Robin *et al.*, 'Multi-Tone Signalling System Employing Quenched Resonators for Use on Noisy Radio-Teleprinter Circuits', *Proc. I.E.E.* **110**, 9, September 1963, pp. 1554-68.
11. A. C. Croisdale, 'Teleprinting Over Long Distance Radio Links II–Automatic Error-Correcting Methods Used Over Long-Distance Radio Links', *Post Office Electrical Engineers Journal* **51**, 1958, pp. 219-25.
    D. A. Chesterman, 'An Automatic Error-Correcting Radio-Telegraph Multiplex System', *Post Office Electrical Engineers Journal* **60**, 3 October 1967, pp. 187-94.
12. *C.C.I.T.T. White Book VII*, Recommendation R31.
13. *C.C.I.T.T. White Book VII*, Recommendation R35.

Chapter 12

Data transmission

12.1 Introduction

With the increasing use of computers and computer-like machines, there is an increasing requirement to transmit information in digital form. There are many eulogies to the better world that this will bring and there are many who predict that, in a few years, the majority of the traffic in the telecommunications system will be digital. The actual range of requirements is very large from a few bits/month for electricity meter reading to several Mbits/s for computer-to-computer communication. There are four easily identifiable types of usage.

(a) Point-to-point transmission–data input to a remote computer–remote control and telemetering.
(b) Data collection–plant monitoring, sales and inventory control etc.
(c) Data dissemination–stock exchange prices, weather reports etc.

and probably the most important for the future,

(d) Interactive stations–air-line booking, on-line computer operation.

For requirements up to 50 bits/s, it is possible to use the Telex system. When a teleprinter system is driven by paper tape, a perforation is used to indicate the condition Z (i.e. stop or mark condition) and for computer applications, a perforation represents a '1'. Hence it is an obvious choice to associate the condition Z with the logical '1'. However, the use of data transmission over a teleprinter circuit needs some restriction, since the system is expecting a start/stop type of transmission.

In the Inland Telex system, the 'line-idle' condition (which corresponds to the state when the teleprinter is in use but not transmitting a character) is signalled by a continuous stop condition, and clearing is produced by a continuous start condition which also indicates that the line is available, i.e. switched on. Hence the data terminal must arrange that during any period in which no data is sent, then a continuous stop (i.e. '1') is sent and also that no code is used which produces more than eight start pulses (i.e. '0') in a row, otherwise the system will clear down.

If the telegraph circuit is likely to meet an automatic regenerator or a t.d.m. system (as it is likely to do on an international call at present and may possibly do on inland calls in the future), then it will be necessary to arrange any data into a start/stop type format, as these signals will be removed by the equipment and automatically reinserted at the remote end.

Also in a Telex call there is equipment which may be actuated by a specific combination of characters (e.g. automatic answer back or operator recall). It is clearly necessary to disable these if data is being transmitted. It has been agreed internationally that any data signals will be sent only after the connection has been made by normal Telex operation and then the sequence SSSS transmitted. This will disable any special equipment and can switch the remote end to data reception. The call can be cleared by the continuous start signal. If the user wishes to revert to teleprinter operation, then it is up to him to find a code combination to achieve this. It will not be possible to re-enable the other equipment.

For a higher dimensional code it is still necessary to use the start/stop 'envelope' for individual 5-bit packages. There is discussion going on about the desirability of a 200 baud telegraph network to allow the use of 7-bit codes.

Ideally all the signals should be carried on an all-digital integrated transmission system whereby all signals are in digital form. There are various proposals for such data networks and some of them are outlined in section 12.4. However, until that day comes the majority of the data signals will have to be carried over analogue channels, most of which will have been designed without any consideration for data transmission.

Analogue data transmission is usually achieved by sending a series of different signals, e.g. discrete voltage levels, in a time sequence. The rate of transmission of signals is limited by the bandwidth and there is a theoretical maximum of $2W$, signals/second for a bandwidth of W Hz, assuming noiseless and distortionless channel [1]. This is called the Nyquist rate. In practice there are many other limitations which will be discussed later.

The maximum number of signal transitions per second is referred to as *bauds*, and if a multi-level signal is used, then the number of bits per second is not the same as the baud rate. Since telegraphy systems are two condition signals then the bits/s rate is equal to the baud rate and this sometimes causes confusion in data systems. For instance, if the bits to be transmitted are taken two at a time and coded into one out of four signals, then the transmission rate will be twice the baud rate.

When noise is taken into account, the fundamental limit is given by Shannon's equation to be

$$C = W \log_2 \left[1 + \frac{S}{N} \right] \text{ bits/s}$$

where S is the average signal power and N is the noise power.

There are a variety of modulation techniques that may be used for data transmission and the choice will depend upon the channel characteristics and the allowable expense of modulator and demodulator. The common techniques are based on amplitude, frequency and phase modulation.

Eye diagram
In any digital link it is desirable to have a measure of how 'good' the system is. A technique which is commonly used is that of the *eye diagram* which gives a graphical indication of the timing and amplitude margin for the detector. This diagram may be produced on an oscilloscope by taking two signalling intervals and superimposing all possible pulse sequences. An example is shown in Figure 12.1(a) for a simple system of unipolar pulses. This signal will be sampled at the mid-point of time and of amplitude, i.e. at the point marked. The size of the 'eye' indicates the margin of error allowed. If the pulses have been distorted then an eye of the form of Figure 12.1(b) will result.

This principle may be extended to multi-level signals and Figure 12.1(c) shows a ternary system where the pulses have been passed through a filter with a raised cosine amplitude response. The effect of noise and phase distortions will close the eye up, making detection more difficult or more liable to error.

The characteristics of a practical channel using modulation may be measured in terms of frequency response and phase or group delay response, but it is not straight forward to convert these to eye closure. One technique used to obtain a direct measurement of the 'goodness' a channel for data use is the *peak-to-average ratio* (PAR) meter[2]. The basis of measurement is to send a series of shaped pulses of carrier frequency along the channel. The effect of frequency and phase distortion will be to distort the pulse and spread its energy, and hence reduce the ratio of the peak signal to the mean power of the test signal. This reduction of the ratio may be used as a measure of the channel performance.

12.2 Use of the telephone network for data transmission

The telephone network was not originally designed with the use of data transmission in mind and hence considerable care must be taken over the matching of a data system to a telephone system. It is worth remembering the characteristics of a telephone circuit as regards their capacity to take data signals[3].

(a) *Overall loss.* With the new switching and transmission plan there should be a maximum of 20 dB loss (at 800 Hz) between any two dependent exchanges in the United Kingdom. Until this plan is in full operation, attenuations of more than 30 dB can be expected on maximally adverse conditions.

(b) *Attenuation/frequency distortion.* The loss at frequencies above and below 800 Hz can be substantially greater from that at 800 Hz. The low frequency end is caused by channel filters and can be 5 dB greater at 400 Hz. The high frequency attenuation is caused by heavily loaded long junction circuits and can be up to 20 dB greater at 2000 Hz (i.e. 50 dB in all) in adverse connections.

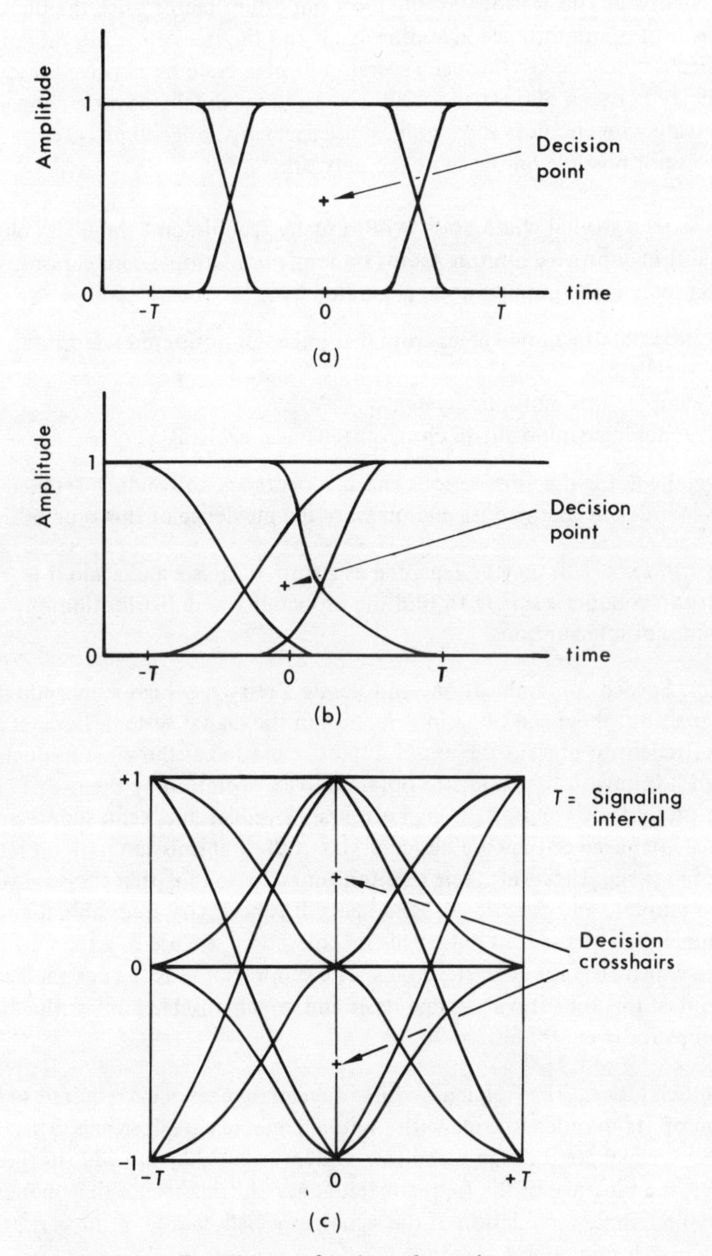

Figure 12.1 Eye diagram for data channels
(a) unipolar pulses ideal reception (b) unipolar pulses with channel distortion (c) ternary pulses

(c) *Group delay*. This is probably the most important factor since the phase response is of no importance in audio signals and the frequency filtering of the audio signal in its passage through several multiplex systems can give up to 2 ms group delay between 800 Hz and 2000 Hz in the maximally adverse connection. On a private wire circuit it is possible to use specially designed phase correctors, but this is not possible on the switched network.

(d) *Noise.* Background white noise is not usually troublesome on inland circuits, although it may prove a limiting factor on long international connections. What is troublesome is the impulse noise generated by:

(i) cross-talk in audio cables from dial pulses or unfiltered teleprinter signals,
(ii) clipping in a multiplex system,
(iii) switching equipment in electromechanical exchanges.

Of these, the last is the most serious and it is necessary to conduct detailed surveys in order to characterise and measure the incidence of this type of noise.

(e) *Interruptions.* This may be regarded as a form of noise, and again it is necessary to conduct a survey to find the probability and distribution of down-times of interruptions.

(f) *Echos.* Speech communication can tolerate a relatively high amplitude of echo signals but these can cause interference in the digital system. In most cases, this is sufficient to prevent the use of duplex operation at the same frequency, although it is obviously possible to obtain duplex operation by the use of different frequency bands. Although echos are troublesome, echo suppressors are unacceptable since they are designed to cut the transmission path on receipt of a speech signal. However, their reaction times are so long that they are very likely to mutilate any data signals, and hence it is necessary to disable them if the channel is to be used for data. This is achieved by the use of a tone at 2100 Hz which lasts for at least 300 ms. The suppressor must be designed so that on receipt of this tone it will disable itself and remain disabled unless the data signal stops for over 100 ms.

(g) *Frequency shift.* The frequency shift of an audio channel is required to be less than ±2 Hz in order to cope with multi-channel telegraph signals. This tolerance is adequate for high speed f.m. systems since they use wide deviations. However, the presence of the frequent frequency shifting means that there will be incidental phase modulation of the signal, especially if any of the carrier links are switched during service.

(h) *In-band signalling equipment.* These are fitted on many trunk lines used in switched connections for supervisory signals (such as clear-down etc.); for instance, in the United Kingdom there are two main systems in the regions

420-900 Hz and 2130-2430 Hz, and in order to remove the possibility of exciting the equipment it is necessary to avoid transmitting energy in these bands. Even in private wire circuits, it may be necessary to take some precaution to avoid operation of this equipment.

Hence there is not much of the spectrum left, and this is one of the reasons the signalling rates are so much less than would be expected.

Modulation techniques

Amplitude modulation is not of much value for use over the switched network. The main reasons are that it is difficult to design an automatic gain control circuit at the receiver to counteract the loss variations. Also the speed of response is slow if an envelope detector is used at the demodulator. This is because it will take several cycles of the carrier signal before the envelope detector can reliably detect the signal.

The other two main choices are frequency or phase modulation. Frequency modulation is the simplest and, for instance, for international use the CCITT recommend [4] the use of f.m. systems for 600 and 1200 baud operation with

	F	F_Z('1', mark)	F_A('0', space)
600 Bauds	1500 Hz	1300 Hz	1700 Hz
1200 Bauds	1700 Hz	1300 Hz	2100 Hz

In addition, an optional return channel, for supervisory and error-control signals, is provided for by a 75 baud channel operating at F_Z = 390 Hz, F_A = 450 Hz. It is found that for a proportion of switched connections, the error-rate of the 1200 baud system is unacceptable, but the 600 baud system will work over the majority of inland dialled connections in the United Kingdom.

It is important to realise the difference between the type of f.m. signal used here and that use, for instance, in the 50 baud system. At 1200 baud the unit element is 833 μs long and this corresponds to only $1\frac{1}{12}$ cycles of a 1300 Hz tone. For this reason the modulators and demodulators for this speed of operation must be made from digital timing circuits rather than *LC* oscillators or discriminators [5].

For applications such as human access to a time-sharing computer, equal signalling rates are required in each direction. The CCITT recommended system for this is Datel 200, and uses f.m. signals in each direction with frequencies [6].

Calling sub.	f_z = 980 Hz	f_A = 1180 Hz
Called sub.	f_z' = 1650 Hz	f_A' = 1850 Hz

It may be shown that phase-modulated systems have the advantage of a higher resistance to white noise interference than have frequency-modulated systems and, in particular, it is possible to use a higher signalling rate over a given bandwidth using phase rather than frequency modulation.

Since an audio channel can suffer many frequency translations with the inherent loss of phase coherency, then it is not possible to use an absolute phase modulation system on the telephone network, but it is possible to use differential phase in which a phase change indicates a '0' and no phase change indicates a '1'. For instance, if the input to the phase modulator is such that a '0' gives one phase and a '1' the other, then the following digital sequence must be recoded before application to the modulator.

Message digit stream	0 1 1 0 0 0 1 0 1
Input to phase modulator	0 1 1 1 0 1 0 0 1 1

This technique obviously requires synchronous detection with some form of bit timing. The latter may be achieved by sending a sequence of '0' at regular intervals to synchronise the remote receiver, or else a smaller phase shift may be imposed on the signal phase shift. The technique can be extended to multi-level phase modulation which is found to be very successful for higher data rates. For instance, 4-level p.m. systems can achieve data rates up to 2400 bauds on the switched network and up to 4800 on private wires[7]. If some form of adaptive equaliser[8] is used at the demodulator then higher rates than these are possible.

Power levels of modulated signals

Since the use of data terminals is effectively uncontrolled, and it is also very likely that their use will become widespread in the future, it is essential that the signals they produce are completely compatible with speech signals as far as any f.d.m. and p.c.m. trunk systems are concerned. The conventional loading of –15 dBm0 makes allowance for about 5 per cent of the speech channels being used for non-speech applications at a fixed power level of –10 dBm0 simultaneous in both directions. However, assuming a higher proportion (10 per cent to 20 per cent) then the maximum power should be reduced to –13 dBm0 in each direction[9].

Since some of the local exchanges may be connected to the trunk network via low loss lines, then the safest assumption is to design on the basis that the 0 dBr point is on the trunk side of the local exchange. Hence the output power of the data set must be adjusted so that the power at the local exchange is not greater than –13 dBm (or –10 dBm if the system is simplex). In order to prevent cross-talk on the local network it is also necessary to specify the maximum power into the line and this is normally 0 dBm.

12.3 High-speed data systems

For higher speed operation than can be provided on conventional telephone channels, it is necessary to use group (48 kHz) or supergroup (480 kHz) frequency bands. For the main trunk routes the f.d.m. system provides a high-grade fixed loss system and hence a form of vestigial side-band a.m. may be used[10]. One of the problems is that it is necessary for the data signal to avoid a group pilot which is situated at 100 kHz in the 60 to 108 kHz group band. It is necessary to get the signal from the subscriber to the multiplex station and this is done by using the local distribution system. The signal is transmitted as simple binary, but with the d.c. component suppressed. Also, at the subscriber's premises is included a scrambler which randomises the input bit stream. Hence repetitive data patterns are removed which would otherwise result in persistent single-frequency components being transmitted on the local network. This speed is usually only available on a private wire basis but the British Post Office have operated a manually-switched 48 kbits/s system on an experimental basis[11].

Higher speeds than 48 kbits/s are possible with a group bandwidth, but the phase distortion at the band edges produces difficulties. This may be avoided by the use of what is called *partial-response coding*[12] which uses signal waveforms with low energy content at the band edges. The use of such waveforms introduces inter-symbol interference and this means that the number of received levels at the sampling instant is greater than the number of levels that are sent. However, this is determinate and may be allowed for by the use of suitable decoders. The effect is to reduce the noise margin, but if the signal-to-noise ratio of the channel is adequate then rates up to 108 kbits/s are possible in effectively 40 kHz of the 48 kHz group bandwidth.

12.4 Low-speed data systems

At the other end of the scale there are many requirements for low data rates, but with very cheap terminals which transmit to a central receiving station over the switched network. For this type of application the use of a parallel transmission system is very economic, since each character to be transmitted is given a combination of tones. The reception equipment is expensive since it must involve a large number of filters together with appropriate detection circuitry. The maximum signalling rate is determined by the necessary recognition time for the received tones, which is of the order of 25 ms, i.e. a signalling rate of 40 bands. This permits character signalling at up to 20 characters/second. if an inter-character rest condition is used. Up to 40 characters/second is possible if no inter-character rest is used and timing information is sent by some means.

The CCITT have standardised a system for this type of use[13] and it is based on 2 or 3 groups of four frequencies in each group. Each character corresponds to one frequency from each group which gives 16 or 64 possibilities for each character (i.e. 4 or 6 bits of information per character). This will cover a large number of the anticipated requirements, and if a larger character set is needed then 8-bit characters (256 combinations) may be sent by sending them in two 4-bit halves and using the third group of frequencies to provide an indication of which is the first and second half of the character.

The recommended frequencies are:

A	920	1000	1080	1160
B	1320	1400	1480	1560
C	1720	1800	1880	1960

The 16 character set uses groups A and C, and if 40 character/second signalling is used, then a selected pair of group B provides timing information. The 256 character set also uses A and C groups and two of the group B frequencies identify the two halves.

It is also possible to use the push-button telephone as a low speed data entry. These normally have twelve buttons and use two groups of four frequencies for their signalling. The voice of frequencies for this purpose is designed to produce a cheap detector and also to minimise the possibility of voice imitation, since the microphone is connected to a line in between pushing the buttons. The frequencies chosen are:

Low band	697	770	852	941
High band	1209	1336	1477	1633

12.5 Data networks

In order to obtain the full benefit from digital transmission it will be necessary to provide an all-digital transmission and switching network with digital access provided at the user's premises. There are many proposals for such systems under active consideration at the present time (1971). The straightforward approach is to provide a subscriber with digital access to a multiplexor at the local exchange using a baseband signalling system, which will work reliably up to 48 kbits/s on normal twisted pair. The multiplexor is arranged to provide whatever transmission rate is requested by the user, and multiplexes the signals into a rate compatible with a p.c.m. transmission system. These then go to digital switching systems to connect required terminals together. This provides a circuit switched digital capability, i.e. a user can set up a digital link between himself and another user at one of the preferred digit rates. The channel itself will be synchronous and the user will have to adapt his equipment to the system.

An alternative technique to circuit switching is *packet switching.* The user assembles the information he wishes to transmit into packets of anything up to 1024 bits and these are transmitted as an entity through the network. This technique allows the use of asynchronous devices at either end and permits the efficient use of trunk capacity as the digital trunk network is only needed whilst information is being passed and not for the duration of the connection time.

REFERENCES

1. The basic theory is covered by W. R. Bennett and J. R. Davey, *Data Transmission*, McGraw Hill, 1965.
2. L. W. Cambell, 'The PAR Meter: Characteristics of a New Voiceband Rating System', *Trans. I.E.E.E. Comm. Tech.* **COM-18**, April 1970, pp. 147-153.

3. For example, M. B. Williams, 'The Characteristics of Telephone Circuits in Relation to Data Transmission', *Post Office Electrical Engineers Journal* **59**, October 1966, p. 151.
P. N. Ridout and P. Rolfe, 'Transmission Measurement of Connections in the Switched Telephone Network', *Post Office Electrical Engineers Journal* **63**, July 1970, pp. 97-104.
4. *C.C.I.T.T. White Book*, Volume VIII, Recommendation V23.
5. L. W. Roberts and N. G. Smith, 'A Modem for the Datel 600 Service–Datel Modem 1A', *Post Office Electrical Engineers Journal* **59**, July 1966, pp. 108-116.
6. *C.C.I.T.T. White Book*, Volume VIII, Recommendation V21. Equipment design described in J. C. Spanton and P.L. Connellan, Modems for the Datel 200 Service, *Post Office Electrical Engineers Journal* **62**, April 1969, pp. 1-10.
7. *C.C.I.T.T. White Book*, Volume III, Recommendation V26. Equipment described in G. W. Adams, 'A Modem for the Datel 2400 Service', *Post Office Electrical Engineers Journal* **62**, 1969, p. 156.
8. C. W. Niessen and D. K. Willim, 'Adaptive Equaliser for Pulse Transmission', *Trans. I.E.E.E. Comm. Tech.* **COM-18**, 4, August 1970, pp. 377-395.
9. *C.C.I.T.T. White Book*, Volume VIII, Recommendation V2.
10. *C.C.I.T.T. White Book*, Volume VIII, Recommendation V35.
11. M. E. Gibson, and D. R. Millard, 'An Experimental Manually-Switched 48 kbit/s Data Transmission Network', *Post Office Electrical Engineers Journal* **64**, 1, April 1971, pp. 52-9.
12. F. K. Beker *et al*, 'A New Signalling Format for Efficient Data Transmission', *B.S.T.J.* **45**, 5, May 1966, pp. 755-8.
13. C.C.I.T.T. White Book, Volume VIII, Recommendation V30.

Chapter 13

Tomorrow's transmission systems

13.1 Introduction

The research effort being spent on millimetre and higher frequency transmission systems has been greatly stimulated by the prospect that such systems offer for accommodating economically the large and increasing volumes of telecommunications traffic foreseen on main trunk routes of the future. Much of this traffic growth will take the form of new services such as picturephone (or viewphone), conference television (confravision) and highspeed data transmission between computer systems. These new services require much greater signal bandwidth than telephony or low-speed data being currently offered and the search for new transmission media has a major objective in the reduction of transmission costs per unit of signal bandwidth.

Another major factor has emerged from studies of the requirements of national networks (both United Kingdom and United States) in the future, which indicate substantial economic, operational and performance advantages in an integrated p.c.m. digital transmission and switching trunk network, as compared with one using conventional f.d.m. analogue techniques. It appears likely that, even without digital switching, it may still be possible to obtain significant performance and operational advantages at no extra cost by using digital rather than analogue transmission.

Since the introduction of p.c.m. digital techniques increases the transmission bandwidth by anything up to 20 times compared with analogue techniques, media offering large inherent bandwidth are required. To make these techniques economical the use of high bit rates to achieve maximum capacity per carrier will be required. This in turn calls for tight specifications on transmission media in terms of attenuation and delay distortion.

The scope for the new high-capacity systems will be conditioned by the growth of traffic arising from existing telecommunication services, i.e. telephony and data, and from the new services such as viewphone and confravision. It is estimated that telephone exchange connections in the United Kingdom will increase four times to a total of about 32 million by the year 2000, and the volume of trunk traffic will increase by a factor of fifteen by the end of the century.

Data terminals operating at speeds up to 6 Mbits/s are estimated to number some 1·5 million by the year 2000, the volume of trunk data being up to 10 percent of the telephony traffic.

The growth of viewphone and confravision are less easily forecasted as they depend upon customer reactions and the cost of the services. If the busy-hour trunk traffic due to these services were the same as present-day telephony traffic, the bandwidth required would be about 100 times greater. This, together with the growth of closed-circuit television for educational purposes, represents a drastic increase in system bandwidth and points to the requirement for capacities of 1 to 10 Gbits/s in the near future.

Systems to fulfil this need will be in the higher frequency bands and they can be broadly classified as guided or radio systems.

13.2 Guided-wave systems

As digital transmission systems spread throughout the network, a number of advantages will be apparent, including:

(a) reduced losses on trunk calls,
(b) better speech quality and lower levels of noise and crosstalk,
(c) ability to provide high-quality music circuits,
(d) reduced errors in data transmission,
(e) ability to accommodate video signals on common transmission plant without significant crosstalk or accumulation of noise on long circuits.

These transmission advantages arise primarily from the ability of p.c.m. digital systems to reject noise, crosstalk and interference below the decision level of the digital signal regenerators.

Two guided-wave systems which currently look like fulfilling the extra capacity requirements are:

(a) Trunk waveguide, TE_{01}-mode shown in Figure 13.1(a)
(b) Optical fibre guide, shown in Figure 13.1(b)

The electrical, mechanical and traffic capacities considered typical of these systems at their present stage of development are indicated in Table 13.1. It should be noted, however, that whereas the trunk waveguide systems are at present well developed to the extent of field trials [1], the optical fibre systems are at a considerably earlier stage of development and much technological improvement is required before feasibility can be established. It will also be noticed from Table 13.1 that a vast expanse of frequency spectrum lies between the two techniques. As yet this barren region has been largely unexploited due to the lack of devices available. It is inconceivable to think of such a large capacity hole being left in the spectrum, and one would imagine much effort to exploit it in the future. At present surface-waveguide techniques look as though they have provided an answer.

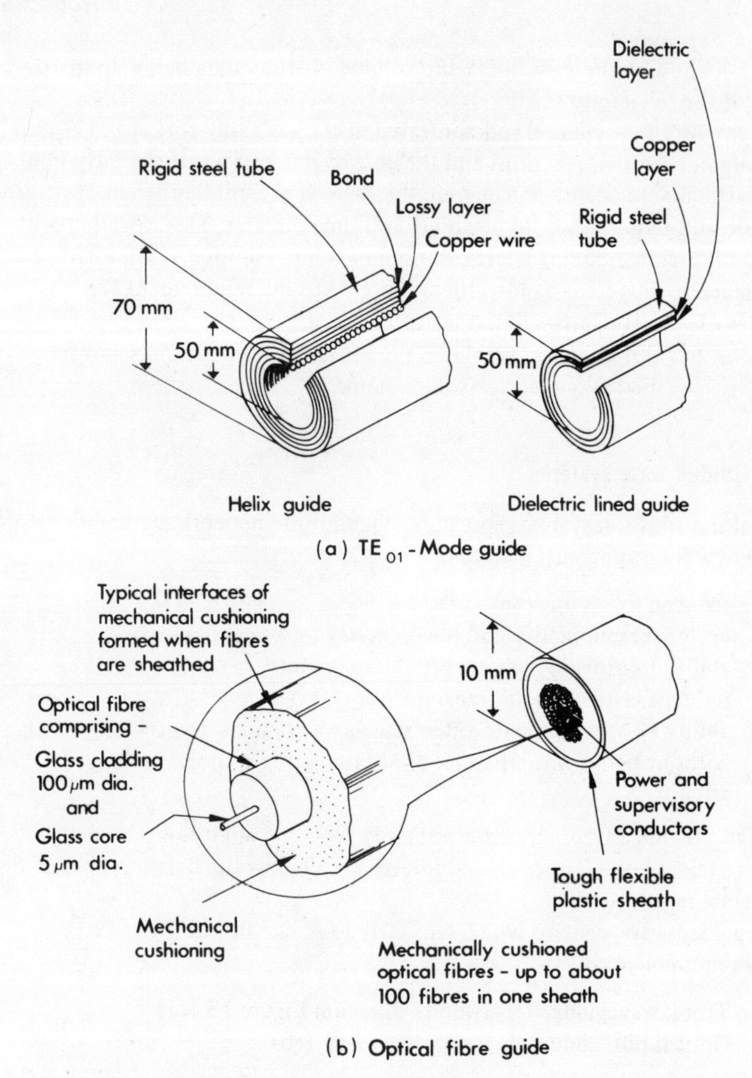

Figure 13.1 TE_{01} and optical guide structures

TE_{01} trunk waveguide system

The unique low-loss quality of the TE_{01} mode helix or dielectric lined cylindrical waveguide enabling repeaters to be spaced some 15-25 km apart is shown in Figure 13.2 This tends to counteract the high cost in providing and laying a transmission medium that demands special considerations in terms of close dimensional tolerances.

·Due to the non-availability of solid-state sources for the repeaters, initial consideration has been given to the lower part of the band below 50 GHz. The primary limitation in this band is the effect of group delay variation with

Table 13.1. Mechanical, electrical and traffic capacities of typical TE_{01}-mode (50 mm) and optical fibre guides

	Characteristic	(a) TE_{01}-mode guide	(b) Optical fibre guide
Mechanical	Diameter	Ext. 70 mm Int. 50 mm	10 mm (100 fibre cable)
	Rigidity	Rigid	Flexible
	Curvature	≮ 100 m radius	≮ 1 m radius
	Laying	Requires special duct	Could use existing duct
Electrical	Frequency range	30 – 120 GHz	3×10^5 GHz approx
	Loss	2·5 – 3·5 dB/km (achievable)	10 – 20 dB/km (objective)
	Repeater spacing	10 – 20 km	1 – 2 km
Traffic capacity	Bit rate	500 Mbits/s (2 Gbits/s)[a] (per carrier)	100 Mbits/s (500 Mbits/s)[a] (per fibre)
	Telephony	400 000	160 000
	Viewphone	5000	2000
	TV (both ways)	250 (fully equipped single guide)	100 (100 fibre cable)

[a] Second stage of development.

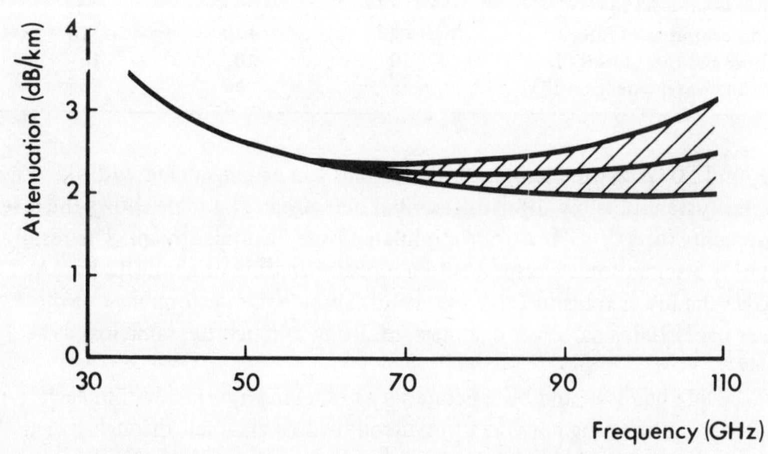

Figure 13.2 Estimated practical waveguide loss limits

frequency. For example, the theoretical slope of group delay at 40 GHz is about 3 ns/km/GHz which, even assuming adequate equalisation, limits the capacity to 600 Mbits/s falling to 500 Mbits/s at 32 GHz. Beyond 50 GHz the attenuation is reasonably constant and of low value and system complexity can be reduced. Due to practical effects of mode-conversion-reconversion and signal distortion [2], losses increase with frequency and an upper limit of operation of the 50 mm pipe is probably around 110 GHz, although reducing the diameter will probably allow operation up to about 300 GHz.

The system envisaged at present is very similar to those considered in Chapter 8 for digital satellite systems, 4-phase p.s.k. being preferred for maximum noise immunity. The basic system lay-out is shown in Figure 13.3. Obviously system levels depend critically on state-of-the-art in solid-state sources. However, using currently available and extrapolated figures, some probable system parameters are summarised in Table 13.2. The permissible waveguide loss at 32 GHz is about 75 dB, indicating a possible repeater spacing of about 21·5 km. At higher frequencies in this band, the reduction of transmitter source power is more than compensated by the reduction of waveguide attenuation.

Table 13.2. System levels for TE_{01} waveguide systems (from ref.[1])

Frequency (GHz)	32	50	110
Bit-rate (Mbits/s)	500	2000	500
Nominal channel bandwidth (GHz)	0·5	2	0·5
kTB (dBm)	−87	−81	−87
Receiver noise factor (dB)	10	12	13
Modulation system	PSK 4-phase coherent	DCPSK 4-phase	PSK 4-phase coherent
Carrier/noise ratio for 1 in 10^9 error rate	16	18·5	16
Margin (dB)	5	5	5
Repeater input level (dBm)	−56	−46	−53
Repeater output level (dBm)	+29	+20	+18
Branching and filter losses (dB)	10	10	16
Permissible waveguide loss (dB)	75	56	55

Beyond 50 GHz, broader bandwidth signals can be considered with the use of a simpler system possibly using differential detection. The table shows the level requirements for a 50 GHz carrier modulated by a 2Gbits/s stream. The resultant waveguide attenuation corresponds to a repeater spacing of about 22 km. At 110 GHz the lower transmitter powers and falling noise performance of the receiver necessitates narrower channels, resulting in much the same loss as at 50 GHz.

Waveguide bands would be separated by branching hybrids. Within each band, filter mutiplexing networks provide individual channels through which each repeater operates.

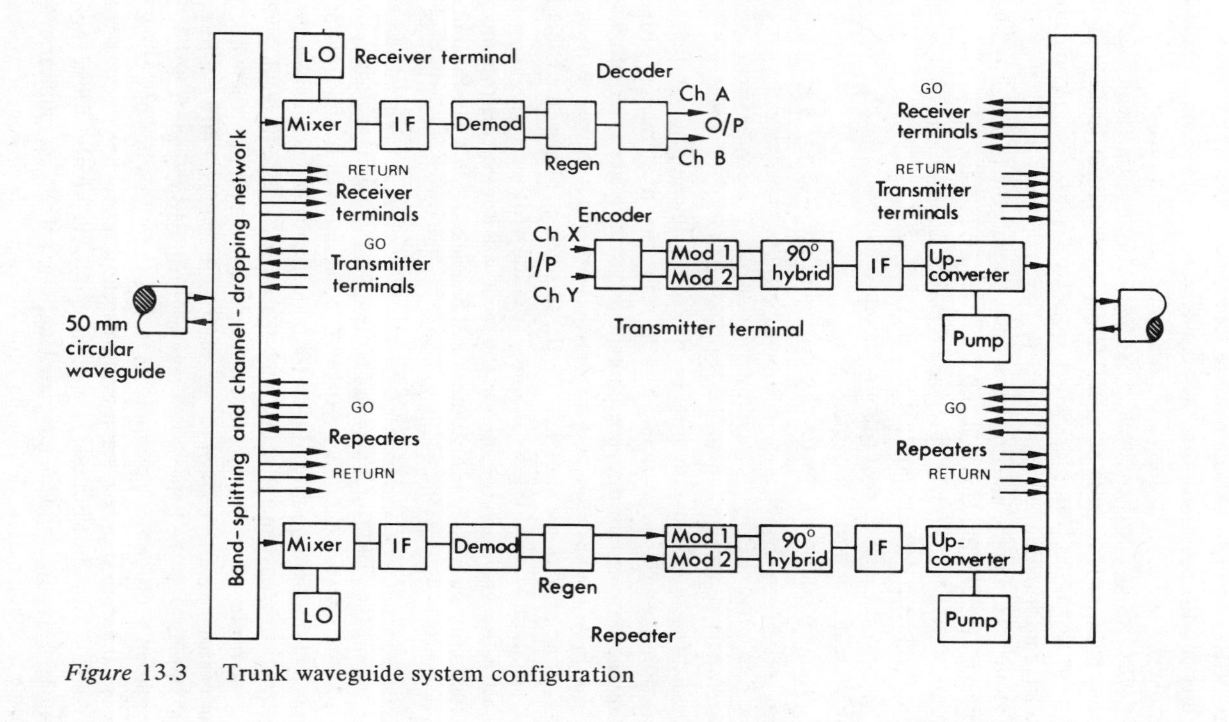

Figure 13.3 Trunk waveguide system configuration

Optical fibre systems

Optical fibre guided-wave systems operating in the near infra-red region (3×10^5 GHz) are of considerable interest, because of their inherent mechanical flexibility and the possibility that they could accommodate economically the traffic likely to be generated on the trunk routes of the future.

Dielectric waveguides have been in existence for some time, and only recently have they been made to work at optical frequencies; one particular kind of guide (see Figure 13.1b) has been proposed as a single mode low-loss transmission line with a large information carrying capacity. The guide takes the form of a glass fibre, about 50 μm in diameter having a small central core of a slightly higher refractive index. The fundamental and operating mode is the HE_{11} which has no lower cut-off. Using dielectric waveguide theory, it is possible to analyse the structure and to design components of a standard type.

At present the most likely communication system using the glass fibre guide would work at near infra-red frequences with a gallium arsenide laser as a transmitter, a modulator imparting some form of p.c.m., and a silicon avalanche diode as a straight detector. The enormous potential bandwidth of such a system is limited to a few GHz by the minimum 4Å line width of the present room temperature lasers. A single mode c.w. gallium arsenide laser would not only greatly increase the bandwidth, but also make the launching more efficient.

The main obstacle at the moment to the realisation of a practical system is the manufacture of a uniform glass fibre guide, virtually free from inhomogenities, and with a low enough loss (about 20 DB/km) to make the repeater spacing economic. So one must wait for the technology of glass manufacture to catch up with the state-of-the-art in communications. Although mode conversion and scattering from inhomogenities in the guide will limit its performance, for extra flexibility ultimately many guides will be packed into a sheath similar to the conventional coaxial cable, and increased bandwidth must favour it as a long term prospect.

At present the main trunk network is composed of f.d.m. analogue coaxial cable and microwave radio-relay systems. Consideration of traffic growth patterns and the trend towards digital techniques already discussed, suggest that a trunk network may emerge with t.d.m. digital TE_{01}-mode waveguides on the back-bone routes between major cities. An extensive supporting network of medium-capacity digital coaxial cables, or, if technical feasibility can be proven, optical fibre cables—the latter would take over trunk routes when they become economic. In addition to its possibilities for use in the junction and trunk networks, the optical fibre system would also find application in the longer term to broadband local distribution from exchanges to the homes of residential customers. This could provide integrated audio/visual telecommunications including telephone, data, viewphone with 'broadcast' television, access to video tape and film libraries, computer memory banks etc. The unique properties of the optical fibre cable, i.e. its flexibility and compactness, ability to provide tens of space-divided broadband channels with simple selection at the receiving end, and its potentially low cost, could make it an attractive transmission medium for such purposes.

13.3 Radio-wave systems

Radio propagation even at the present microwave network frequencies is affected to some extent by the weather. At the higher frequencies now under consideration for future systems, the effect of the weather may be much more drastic.

When rainfall is present in the radio path, there will be substantial attenuation of the signal, necessitating repeater spacing to be reduced to something like 5 km, in addition to fading caused by refraction. The wavelength at the higher frequencies is no longer very large compared with the diameter of rain drops which both scatter and absorb the energy of the radio wave.

To plan radio-relay systems which may be affected by weather, it is necessary to know how frequently rainfall of a given intensity is likely to occur in a particular area, how far it extends and how long it will last. Intensity is important because the heavier the rain, the larger the raindrops, and the greater the amount of radio energy scattered and absorbed. Many investigations both on terrestrial and satellite links must be made before these systems can be planned for a given reliability.

Terrestrial systems

In view of the increasing attenuation with frequency characteristics of the atmosphere and the traffic growth predictions of the future, radiating systems would seem to have a fairly limited usefulness. However, in the near future they will have a part to play in the network, probably in the form of 'pole-line' systems with completely integrated self-powered repeaters placed on the top of simple towers and spaced about 5 km apart. The reduction in cost of the repeater equipment, brought about by advances in microwave integrated circuitry and solid-state technology, make this extremely attractive economically. The somewhat restricted bandwidth could be increased by techniques such as frequency re-use on orthogonally polarised signals and the use of optimised modulation techniques. Because of the cell-like structure of areas of rainfall it could be possible to combat the rapid fades and 'drop-outs' caused by really heavy rainfall by route diversity techniques. One form of this is to operate two paths in parallel at such a spacing so that one of the routes avoids the offending rainfall. Exactly how the rainfall is likely to effect high speed digital signals and degrade error-rates is something which remains unanswered to date.

The determination of the relationship between attenuation and rainfall rate is fundamental to an understanding of the effect of rainfall in radio propagation, and this has to be achieved before systems are installed. Such short-hop radio systems are likely to find application to the junction network and on spur routes, rather than inter-city trunk routes.

There is also a possibility of utilising radio systems for urban distribution of telephony and wideband data and video to the subscribers premises. This system would make use of the oxygen absorption at around 60 GHz to limit the 'broadcast' from telephone exchanges, situated ideally at the centre of the subscriber area, so that adjacent areas did not interfere. The system is very attractive to the operating authorities who have an increasingly large amount of their capital bound up in buried cables. There are, however, many problems of instigating such a system, perhaps the most severe of which is the cost of the subscribers equipment to enable him to transmit and receive signals to the exchange. Advances in equipment technology will prove whether this is a feasible system and whether it will be a serious rival to the fibre-optical cable system.

Satellite systems

Many of the restrictions on higher frequency radio-relay transmissions apply equally to satellite systems. The main difference here is that the propagation path traverses the troposphere only over a small fraction of its length. However, the effects of precipitation may be just as marked, as the path is a slant one and rain clouds have an increased distribution towards the ceiling of the troposphere.

However, the flexibility of the satellite system is only just being realised with the introduction of domestic systems integrating telephony, television and data. The distribution of television signals from programme centres to broadcasting stations involves a great deal of terrestrial network capacity in most countries. These networks could be replaced by a single emission from a satellite for each programme, fed from an earth station associated with the programme centre and received at an earth station located near to each broadcasting station. The distribution of television signals for entertainment and education to a large number of wired-broadcasting networks has already been planned. Other applications could include two-way high capacity data links forming a users' network, or the transmission of high-speed facsimile signals for newspaper reproduction at dispersed printing plants. Ultimately these developments may lead to satellite television broadcasting direct to the home, the receiver unit possibly being adapted to receive facsimile signals of newspaper sheets from any part of the world.

Another new field of application for satellite communications will be the mobile services, linking aircraft and ships to base. With the advent of the supersonic airliner, this will provide a valuable aid to air-traffic control in regular status up-dating to ensure low collision rates in our busy air lanes. The existing h.f. propagation is bad, and danger of collision is significantly increasing. A few systems have already been proposed using f.m. at carrier frequencies of around 1·5 GHz, where the main problem appears to be in the design of satisfactory aircraft equipment. A really urgent need for reliable trans-oceanic aircraft communications will emerge very shortly and satellite communications is the only foreseeable means of providing it.

REFERENCES

1. *I.E.E. Conference* Digest on Trunk Telecommunications by Guided Waves, London 1970.
2. A. E. Karbowiak, *Trunk Waveguide Communications.* Chapman and Hall, 1965.

Appendixes

A. THE CONCEPT OF BALANCE RETURN LOSS

The concept of balance return loss may be understood more clearly by considering a detailed analysis of a hybrid transformer.

If the input impedance of the GO amplifier and the output impedance of the RETURN amplifier are each equal to R and the line and balance impedances are given by

$$Z = aR \qquad \text{(line impedance)}$$
$$N = bR \qquad \text{(balance impedance)}$$

then an equivalent circuit of the hybrid transformer is that shown in Figure A1, where V is the output voltage of the RETURN amplifier (a and b will in general be complex quantities).

If the transformers are assumed to be ideal, then the voltages across each winding of a transformer will be proportional to the turns ratio as indicated on the diagram. The currents and voltages are related:

$$i_0 = \frac{v - v_0}{R} \qquad i_1 = \frac{n(v_0 - v_1)}{aR}$$

$$i_2 = \frac{n(v_0 + v_1)}{bR} \qquad i_3 = \frac{v_3}{R}$$

Since the transformers are assured ideal, then the sum of all the ampere-turns on each transformer must be zero, i.e. for the currents shown:

$$i_0 = n(i_1 + i_2)$$

$$i_3 = n(i_1 - i_2)$$

By substituting the first four equations into the last two it is possible to obtain two equations containing v_0 and v_3 only as unknown, which may then be found. For the normal case $n = 1\sqrt{2}$, and in this case

$$v_3 = \frac{\frac{1}{2}(b - a)}{(1 + a)(1 + b)} \text{ V}$$

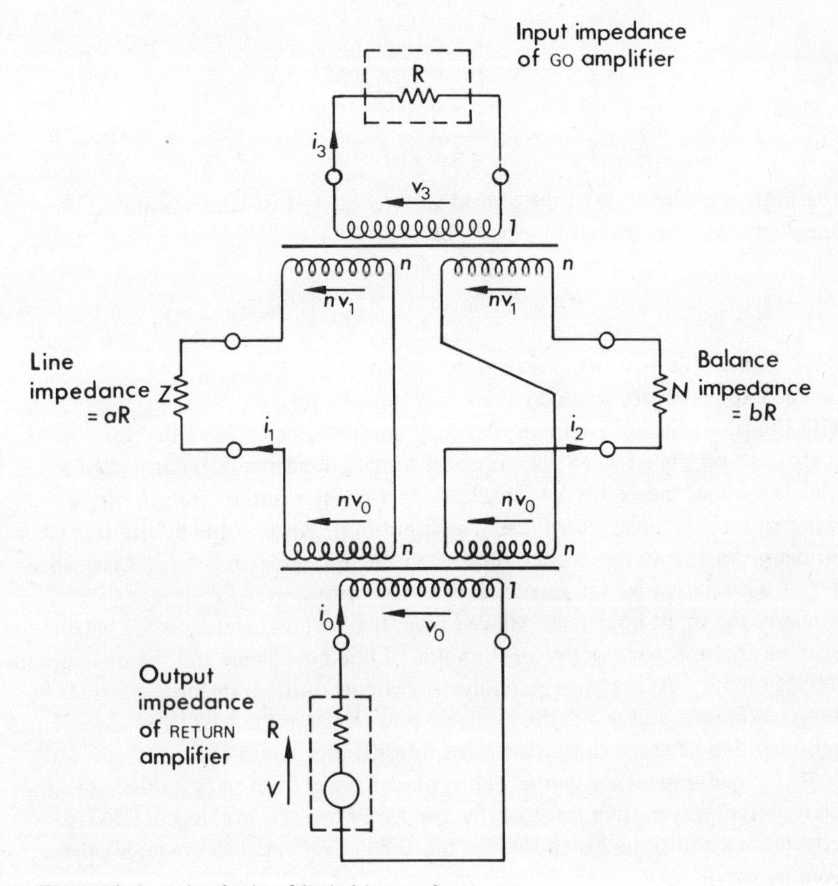

Figure A.1 Analysis of hybrid transformer

which may be rewritten

$$v_3 = \tfrac{1}{4}\left[\frac{1-a}{1+a} - \frac{1-b}{1+b}\right] V$$

The quantity of practical interest is the trans-hybrid insertion loss, which is the ratio of the power in the input resistance of the GO amplifier produced by the RETURN amplifier when it is connected directly to the power delivered when the hybrid transformer is inserted. For a direct connection the voltage across the input resistance is $v_3 = \tfrac{1}{2}$ V, hence the insertion power loss ratio is given by:

$$10 \log_{10} \left|\frac{v_3'}{v_3}\right|^2 = -20 \log_{10} \tfrac{1}{2}\left|\frac{1-a}{1+a} - \frac{1-b}{1+b}\right| \text{ dB}$$

$$= 6 + 20 \log_{10} \left|\frac{1}{\rho_L - \rho_N}\right| \text{ dB}$$

where $$\rho_L = \frac{1-a}{1+a} = \frac{R-Z}{R+Z}$$

and $$\rho_N = \frac{1-b}{1+b} = \frac{R-N}{R+N}$$

The 6 dB is the loss due to the hybrid and hence the loss which is due to the impedance mismatch is given exactly by

$$B_S = 20 \log_{10} \left| \frac{1}{\rho_L - \rho_N} \right| \text{ dB}$$

If $a = b$ then this quantity is seen to be infinite.

The expression for ρ_L and ρ_N are in a form familiar to those acquainted with the transmission-line theory, as they are the value of the reflection coefficient produced when a source of internal impedance R is connected to a load of impedance Z (or N). This leads to a physical interpretation of the expression for balance return loss. The input impedance of the hybrid transformer viewed from the port connected to the line is R (for $n = 1/\sqrt{2}$) irrespective of the value of N. An open circuit voltage source of $V/\sqrt{2}$ is also seen. Similarly the input impedance viewed from the port connected to the balance network is also R irrespective of the value of line impedance and the open-circuit voltage seen is $V/\sqrt{2}$. Hence an equivalent circuit of the transformer is that shown in Figure A.2(a), i.e. the available power† from the RETURN amplifier is split into two halves with generators of internal impedance R.

If the load impedance connected to one of the generators is not equal to the internal impedance, then some of the available power from the generator is reflected back into the generator and the ratio of reflected to available power is given by ρ_L or ρ_N.

As has been explained, any power coming from the line will split into two and will be dissipated equally in the input impedance of the GO amplifier and the output impedance of the RETURN amplifier. Similarly any power coming from the balance network will be split equally between these two impedances. The reflected power coming from the line or the balance network will be treated identically and will also split into two halves. Hence the power flow in a hybrid is as shown in Figure A.2.

Power from the RETURN port will split equally into the LINE and BALANCE ports. At each port some of the power will be reflected back with a voltage ratio of ρ_L and ρ_N and each of the reflected powers will then split equally between the GO and RETURN ports. Because of the cross-connections on the transformer the reflected power will subtract, and hence if the reflection coefficients are equal (i.e. if $N = Z$) there will be no net power exiting from the GO port. If the reflection coefficients are not equal then the portion of insertion

† Available power is defined as the maximum power that can be obtained from a generator, i.e. the power delivered to a matched load.

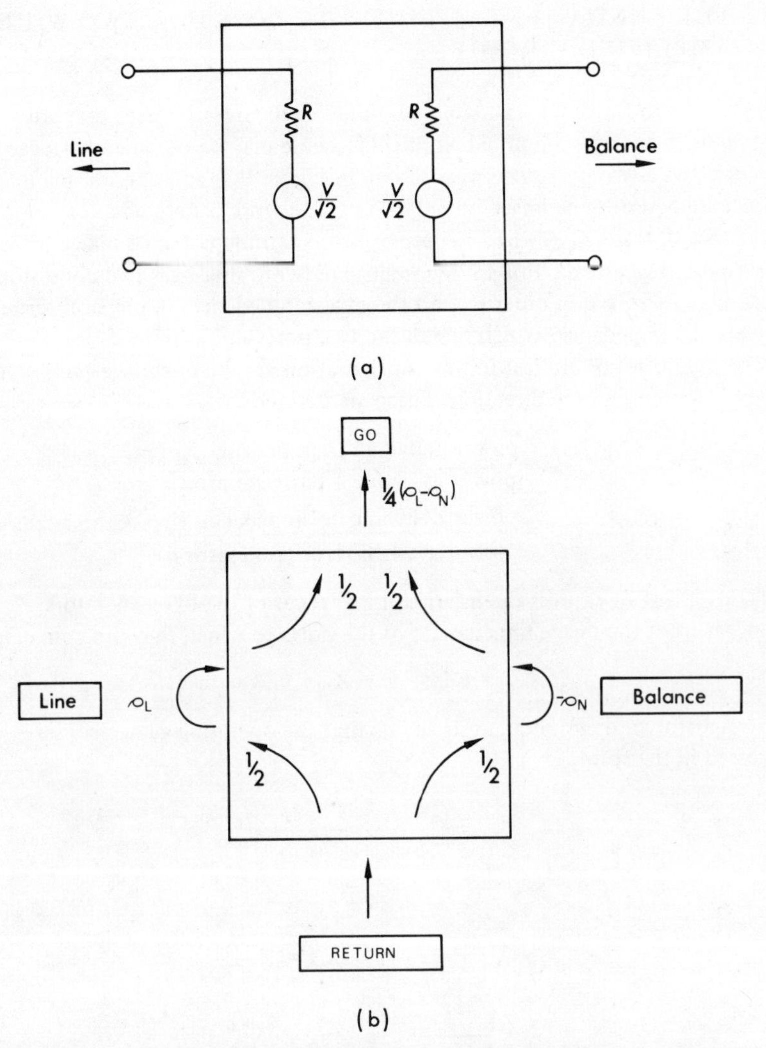

Figure A.2 (a) equivalent circuit of hybrid transformer with respect to RETURN Amplifier (b) power flow in hybrid transformer

loss due to the mismatch will be the difference between the reflection coefficients, which is the expression obtained above.

If ρ_N is small (i.e. b near to unity) then the approximate expression used for $\rho_L - \rho_N$ is

$$\rho_L - \rho_N = \frac{1-a}{1+a} - \frac{1-b}{1+b} \simeq \frac{b-a}{b+a}$$

this is exact for $b = 1$ and is zero for $b = a$. For b near to unity it is still found to be an adequate approximation.

B. THE FUNDAMENTAL LIMITS TO LOSS IN A TWO-WIRE AMPLIFIED CIRCUIT

It is pqssible to find some fundamental limits to the overall losses that are obtainable in a 2-wire amplified circuit[1]. These may be obtained by regarding the circuit as a two-port system, as shown in Figure B.1, and considering its image parameters. (The image impedance of a network is defined as the input impedance of port, Z_{01}, when the other port is terminated in its image impedance Z_{02} and vice-versa. For an asymmetric network this gives two conditions which may be solved in order to find the image impedance. In physical terms they are the impedances which match the two ports).

The quantities of interest in this configuration are the *operating gains* of the two-port system in each direction. These are defined as:

$$G_{21} = \frac{\text{power delivered to impedance } Z_b}{\text{power available from generator } a}$$

$$G_{12} = \frac{\text{power delivered to impedance } Z_a}{\text{power available from generator } b}$$

It is convenient to express the internal impedances of the two generators, Z_a and Z_b, in terms of the image impedances of the ports to which they are connected:

$$Z_a = aZ_{01} \qquad Z_b = bZ_{02}$$

Straightforward circuit analysis shows that the operating gains may be expressed in the form:

$$G_{21} = \frac{(Z_{21}/Z_{12})G_o M_a M_b}{(1 - G_o \rho_a \rho_b)^2}$$

where

$$M_a = \frac{4a}{(1 + a)^2}, \qquad M_b = \frac{4b}{(1 + b)^2},$$

$$\rho_a = \frac{1 - a}{1 + a}, \qquad \rho_b = \frac{1 - b}{1 + b}$$

$$G_o = (G_{12}' G_{21}')$$

and G_{12}' and G_{21}' are the operating gains under matched conditions, i.e. $Z_{01} = Z_a^*, Z_{02} = Z_{10}^*$.

As explained by Llewellyn there are convenient physical interpretations of these various quantities. The M_a term is due to the mismatch between a generator of internal impedance Z_a and an image impedance of Z_{01}. It is the ratio of the power delivered to an impedance Z_{01} to the available power from the generator and has a maximum value of unity when the image impedance is matched to the generator impedance. The ρ_a factor is the reflection factor at

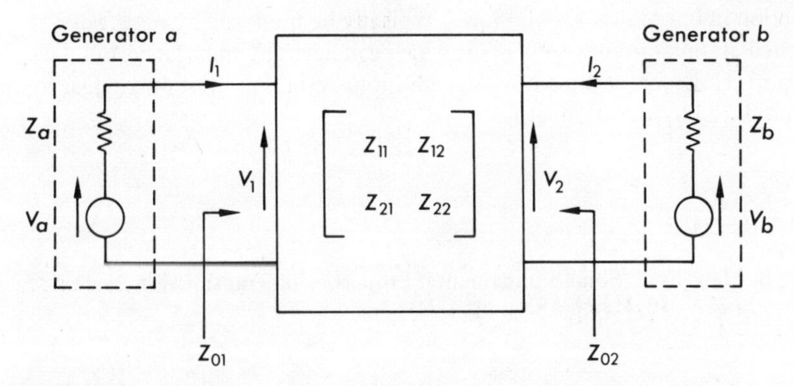

Figure B.1 Linear two-port system with terminations. Z_{01}, Z_{02} are image impedances, Z_{11} etc. are z parameters, i.e. $V_1 = Z_{11} I_1 + Z_{12} I_2$ $V_2 = Z_{21} I_1 + Z_{22} I_2$

this junction. Note that $M_a + |\rho_a|^2 = 1$ which simply states that the sum of the delivered power and the reflected power is equal to the available power.

The G_o term is called the image gain and is the geometric mean of the two operating gains when both the generators are matched to the image impedances. It may be shown that the product $G_{12}G_{21}$ has its maximum value under these conditions.

The important fact to note about the expression for G_{21} is that the factor in its denominator, $1 - G_o \rho_a \rho_b$. This factor can go to zero under certain conditions and this corresponds to infinite operating gain, i.e. instability (the term is in fact of a similar form to the $1 - \mu\beta$ in the denominator of the equation for the closed-loop gain of an amplifier with an open-loop gain of μ and a feedback ratio of β; the square occurs because we are considering power ratios). Consideration of this factor will indicate conditions under which instability is likely to occur. If a two-port is required to operate stably under all terminal conditions then the maximum magnitude of the reflection coefficients ρ_a and ρ_b will be unity and will occur when Z_a and Z_b are pure reactances (which include open and short circuits) and the image impedances are pure resistances, i.e. *a* and *b* are both purely imaginary. Since the reflection coefficients could have any phase angle, then in order to guarantee stability for all combinations of terminations $|G_o| < 1$ since $|\rho_a \rho_b|$ could be equal to unity with particular combinations of terminations. This is the basis of the general rule stated earlier, that for a two-port circuit to be stable under all combinations of terminations the magnitude of the image gain must be less than unity; hence $|G_{12}' G_{21}'| < 1$. Finally, since it may be shown that $G_{12} G_{21} < G_{12}' G_{21}'$, then $|G_{21} G_{21}| < 1$ for stability, i.e. the product of the magnitude of the operating gains must be less than unity or expressed alternatively, and the sum of the operating losses expressed in dB must be greater than zero.

If there is some restriction upon the range of generator impedances then a greater value of $|G_o|$ is permissible whilst still guaranteeing stability. This reduction in impedance variation may typically be produced by means of a matched attenuator between a point which can have any impedance and the two-port. However, in these cases the additional gain is at most only equal to the loss produced by the attenuator.

REFERENCE

1. F. B. Llewellyn, 'Some Fundamental Properties of Transmission Systems', *Proc. I.R.E.* **40**, March 1952, pp. 271-83.

C. DERIVATION OF THE F.M. IMPROVEMENT FACTOR

Consider an f.m. spectrum containing a noise power (spectral noise density) of kT watts/Hz of bandwidth.

The noise voltage referred to a 1 Ω resistor is thus $\sqrt{(kT)}$. Let us consider firstly the case of a single interfering sinusoid of this value. Also take the carrier power to be C watts (voltage $\sqrt{C}$). The noise voltage, being much less than the carrier, will phase modulate the carrier and the maximum phase modulation will be as shown in Figure C.1.

$$\theta = \tan\left(\frac{kT}{C}\right)^{1/2} \simeq \left(\frac{kT}{C}\right)^{1/2} \text{ radians}$$

for small values of the argument.

The instantaneous phase modulation ϕ is

$$\phi = \left(\frac{kT}{C}\right)^{1/2} \sin(\omega_n t + \theta_n) \text{ radians}$$

whilst the instantaneous frequency deviation, f_i, is

$$f_i = \frac{d\phi}{dt} = \left(\frac{kT}{C}\right)^{1/2} f_n \cos(\omega_n t + \theta_n) \text{ Hz}$$

Hence the peak frequency deviation, f_d, is

$$f_d = \left(\frac{kT}{C}\right)^{1/2} f_n \text{ Hz}$$

and the r.m.s. frequency deviation f_{rms} is

$$f_{\text{rms}} = \left(\frac{kT}{2C}\right)^{1/2} f_n \text{ Hz}$$

The r.m.s. output power after detection is proportional to the output voltage squared and thus to the r.m.s. frequency deviation squared.

Of practical interest are the effects of a band of random noise about a carrier,

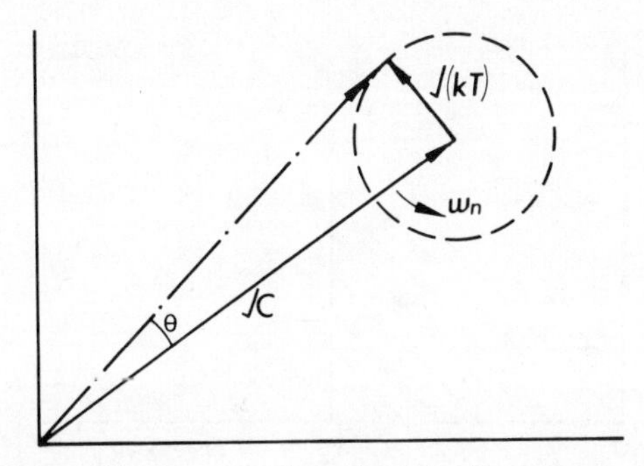

Figure C.1 Phasor diagram—carrier plus interfering sinusoid

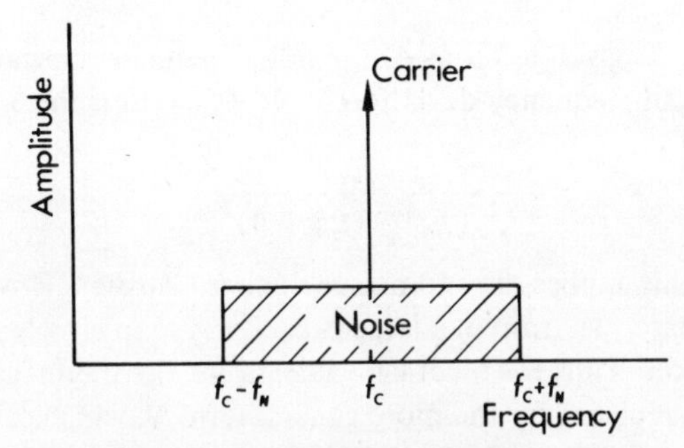

Figure C.2 Flat random noise added to carrier

i.e. (a) the total noise appearing in the baseband of a television signal, and (b) the noise in a particular slot (for example, the noisiest channel in a telephone multiplex group).

The random noise can be considered to consist of a large number of sinusoids, of equal amplitude and arbitrary phase (Figure C.2). The analysis given so far on a per-Hertz basis may be applied as a band of noise N Hertz wide and can be thought of as equivalent to N approximately uniformly-spaced sinusoids. Then, if it is assumed that the peak noise is much less than the carrier, the noise power may be obtained by scanning the N sinusoidal terms across the band. As already seen, the baseband noise/Hz in an f.m. system will vary directly with f_n, and therefore increase linearly with baseband frequency (Figure C.3). This is the so-called triangular noise spectrum of an f.m. system and explains why one considers the baseband top telephone channel to have the worst noise performance.

The r.m.s. noise power in the top 4 kHz telephone channel in the baseband at the f.m. detector output is then

$$N_p = 2\int_{f_m - 4}^{f_m} \left(\frac{kT}{2C}\right) f_n^{\,2}\, df_n = \frac{kT}{3C}\,\{f_m^{\,3} - (f_m - 4)^3\}$$

where f_m is the top channel modulating frequency in kHz.

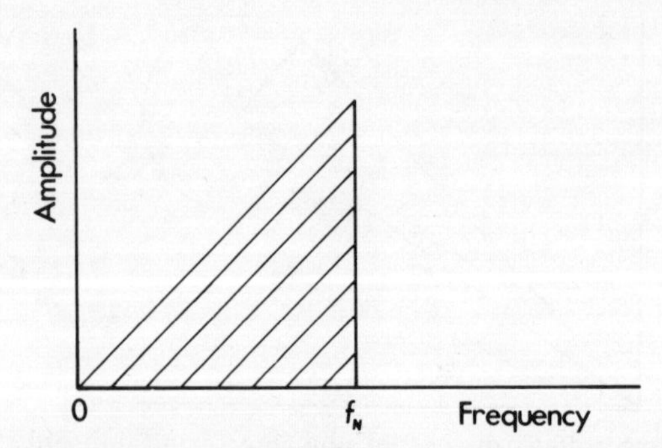

Figure C.3 Triangular baseband noise

The noise output is given in terms of the noise frequency deviation squared. The signal in terms of frequency deviation in $S \propto f^2_{rms}$, thus the S/N is

$$\frac{S}{N} = \frac{3C}{kT} \cdot \frac{f^2_{rms}}{\{f_m{}^3 - (f_m - 4)^3\}}$$

Strictly speaking, this applies only to noise which is Gaussian. However, the intermodulation noise spectrum in a large capacity system closely resembles the Gaussian distribution. Thus for practical calculations the thermal spectral noise density kT may be replaced by the more general term N, which can either be thermal noise, intermodulation noise or both combined. Taking thermal noise $N = kT\,B_{rf}$ we have

$$\frac{S}{N} = 3\,\frac{C}{N\{f_m{}^3 - (f_m - 4)^3\}}\,f^2_{rms}\,B_{rf}$$

and using the approximation $f_m{}^3 - (f_m - 4)^3 \doteqdot 3 \times 4 \times f_m{}^2$,

$$\frac{S}{N} = \frac{C}{N}\,\frac{f^2_{rms} B_{rf}}{4 f_m{}^2}$$

or

$$\frac{S}{N} = \frac{C}{N} + 20 \log\left(\frac{f_{rms}}{f_m}\right) + 10 \log\left(\frac{B_{rf}}{4}\right)$$

where B_{rf} is in kHz.

Thus the advantage over an a.m. system which has a linear S/N to C/N characteristic is

$$I = 20 \log\left(\frac{f_{rms}}{f_m}\right) + 10 \log\left(\frac{B_{rf}}{4}\right)$$

The first part expresses the actual modulation advantage and the second the bandwidth advantage due to the fact that the 4 kHz channel does not occupy the total r.f. noise bandwidth.

N.B. Unless the peak frequency deviation is equal to or greater than $\sqrt{2}$ times the frequency of the top telephone channel, the actual f.m. advantage is negative.

D. EARTH STATION NOISE TEMPERATURE

A satellite receiving station consists of an antenna and a receiver: let T_a be the antenna temperature due to the sky noise etc., and T_R the effective input noise temperature of the receiver. The latter includes noise contributions due to cascade amplifiers and to the loss of the feeder. The sum of the temperatures is the total effective system noise temperature of operating temperature T_S (this is the one referred to in the G/T ratio)

$$T_S = T_a + T_R \tag{D.1}$$

and gives the total system noise temperature referred to the input of the receiver.

(a) *Noise temperature of cascaded networks* Consider two networks in tandem as shown in Figure D.1, having effective input temperatures T_1 and T_2 and available gains G_1 and G_2. The tandem connection is connected to a noise source having a noise temperature T.

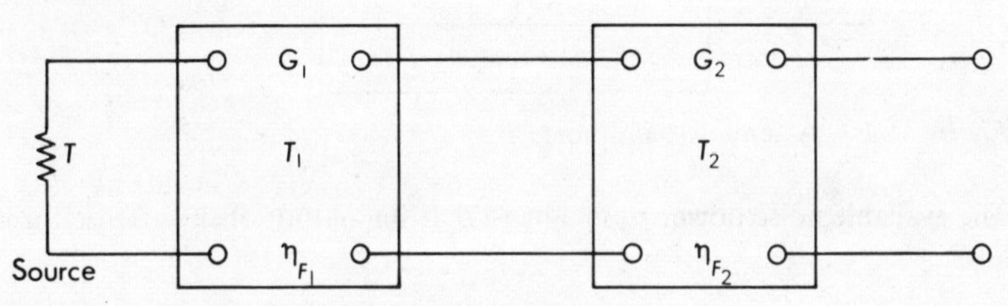

Figure D.1 Cascaded networks and noise

At the output, the following noise powers appear in a small bandwidth ΔB.

(i) due to the noise source, $N_{SO} = G_1 G_2\, kT\, \Delta B$;
(ii) due to noise in network, 1 $N_{10} = G_1 G_2\, kT_1 \Delta B$;
(iii) due to noise in network, 2 $N_{20} = G_2\, kT_2\, \Delta B$;
(iv) total noise at output, $N_{OT} = (N_{SO} + N_{10} + N_{20})$
$= kG_2(G_1 T + G_1 T_1 + T_2)\Delta B$;
(v) Output noise due to networks alone – $N_{O12} = kG_2(G_1 T_1 + T_2)\Delta B$

The effective input temperature of the two networks in tandem is then by definition (see Chapter 4)

$$T_{12} = \frac{kG_2\,(G_1 T_1 + T_2)\Delta B}{G_1\, G_2\, k\Delta B} = T_1 + \frac{T_2}{G_1}$$

This result can easily be generalised to n networks in tandem to give

$$T_{\text{in}} = T_1 + \frac{T_2}{G_1} + \frac{T_3}{G_1\, G_2} + \cdots + \frac{T_n}{G_1\, G_2 \ldots G_{n-1}} \tag{D.2}$$

and using the relationship between effective input noise temperature and noise factor η_F,

$$T = T_o(\eta_F - 1) \tag{D.3}$$

$$\eta_{F_{\text{in}}} = \eta_{F_1} + \frac{\eta_{F_2} - 1}{G_1} + \cdots + \frac{\eta_{F_{n-1}}}{G_1 G_2 \ldots G_{n-1}} \tag{D.4}$$

(b) *Noise temperature due to an attenuator* Consider an attenuator inserted between a noise source (T) and a load as shown in Figure D.2.

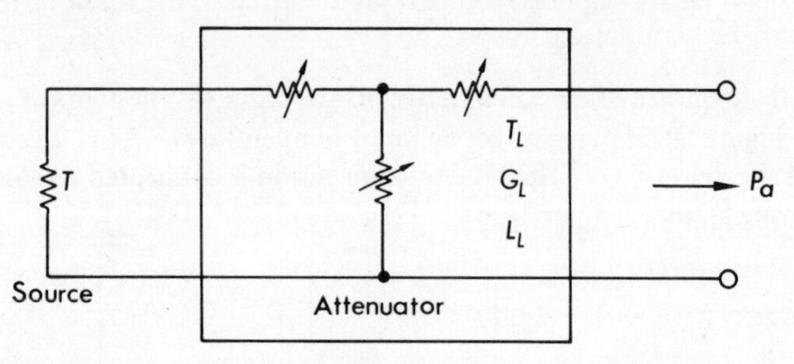

Figure D.2 Attenuator and noise

The available noise power p_a in a band B at the output of the attenuator is as follows:

(i) if $G_L = 1$ (0 dB), then $p_a = kTB$ watts
(ii) if $G_L = 0$ (infinite loss), then $p_a = KT_LB$ watts, and is due to the attenuator loss.

This difference in output powers with attenuator loss is due to the difference in temperature between the noise source and attenuator ($T - T_L$), and this may be regarded as the 'signal' produced by the source which is then attenuated by the attenuator loss. The 'signal' power at the output of the attenuator is then

$$G_L kB(T - T_L) \qquad 0 \leqslant G_L \leqslant 1$$

The total power output of the attenuator is the sum of the 'signal' and noise

$$p_a = G_L kB(T - T_L) + kBT_L$$
$$= kB[G_L T + (1 - G_L)T_L]$$

The effective input noise temperature of the attenuator is then

$$T_{L_e} = \frac{p_a}{kBG_L} - T = \frac{kB[G_L T + (1 - G_L)T_L]}{kBG_L} - T$$

$$= \frac{1 - G_L}{G_L} T$$

The loss is usually given in dB as $L = 10 \log_{10} (1/\alpha)$ dB and so we call the gain ratios α. Hence

$$T_{L_e} = \left[\frac{1}{\alpha} - 1\right] T \tag{D.5}$$

The noise factor of the attenuator is

$$\eta_F = 1 + \frac{T_L}{T_o}\left(\frac{1}{\alpha} - 1\right)$$

It is important to realise that this result only applies for attenuating networks that achieve the attenuation through lossy elements such as resistors.

(c) *Earth Station temperature* Consider the earth station receiver shown in Figure D.3. The effective input noise temperature of the receiver can be calculated using (D.2), noting that the first stage amplifier is high gain and so low-loss contributions from other than the first two stages are negligible, and using (D.3) to express the noise of the second stage amplifier, which being larger is usually quoted as a noise factor,

$$T_R = T_1 + \frac{T_o(\eta_{F_2} - 1)}{G_1}$$

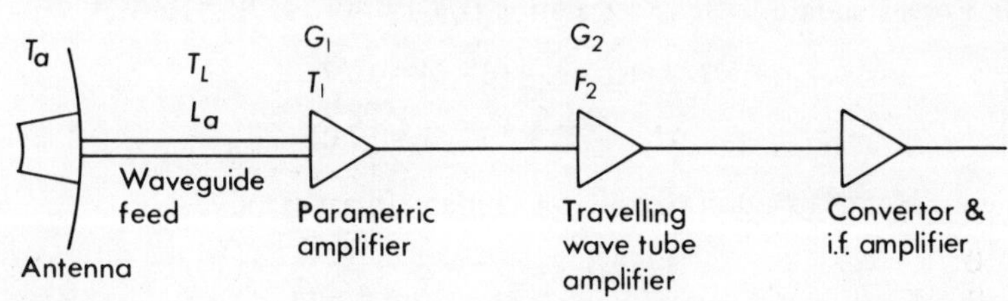

Figure D.3 Earth-station receiving configuration

Equation (D.5) may be used to calculate the effective input temperature of the waveguide as

$$T_{L_c} = T_L\left[\frac{1}{\alpha} - 1\right]$$

where $L_a = 10 \log_{10} (1/\alpha)$, N.B. T_L is usually taken as 290 K.

Hence the total effective temperature of the input to the waveguide is

$$T_a + T_L\left[\frac{1}{\alpha} - 1\right]$$

and referring this through the lossy line to the receiver input,

$$\alpha T_a + T_L(1 - \alpha)$$

Hence using equation (D.1) the earth station receiver temperature is

$$T_S = \alpha T_a + T_L\,(1-\alpha) + T_1 + \frac{T_o\,(\eta_{F_2} - 1)}{G_1} \tag{D.6}$$

(d) *Example* Typical earth station noise contributions are as follows:

T_a – { spillover noise 13 K; sky noise, clear weather 23·6 K; reflector surface noise 1·4 K } 38 K
T_L – feeder temperature ambient 290 K
L_a – feeder loss (includes diplexer) 0·3 dB
T_1 – cooled parametric amplifier 10 K
G_1 – gain of paramp 26 dB (400)
N_{F_2} – travelling wave tube noise factor 8 dB (6·3)

$$0{\cdot}3\text{ dB} = 10\log_{10}(1/\alpha)$$

$$\alpha = 0{\cdot}94$$

Whence, using equation (D.6),

$$T_S = 0{\cdot}94(38) + 290(1 - 0{\cdot}94) + 10 + \frac{290(6{\cdot}3 - 1)}{400} = 67{\cdot}4\text{ K}$$

N.B. For a standard INTELSAT station $G/T_S = 40{\cdot}7$ dB/K. Hence

$$G - 10\log_{10}(67{\cdot}4) = 40{\cdot}7$$

$$G = 59\text{ dB}$$

which corresponds to a dish size of 26 m for an efficiency of 75%.

E. DIGITAL SYSTEM CAPACITY EQUATIONS

(a) *P.c.m./p.s.k./f.d.m.a. system*
Due to voice switching in the SPADE system, 40 per cent of the slots are utilised at any instant.

The down-path thermal noise/transponder is thus ($kT_D \times 36 \times 0{\cdot}4 \times 10^6$) watts and the down-path equation is

$$\frac{C}{N_D} = (E_S - b\,o + \frac{G}{T} - 10\log_{10}(k\,.\,14\,.\,4 \times 10^6)$$

Optimum back-off may be found from Figure 8.9 in the same way as for the f.m.–f.d.m.a. case adding 3 dB to the ∞ curve this time.

80 per cent of the noise is attributed to the down-path, and thus the operating C/N is found from

$$\frac{C}{N_T} = \frac{C}{N_D} - 10 \log \left(\frac{10}{8}\right)$$

Characteristics of 4ϕ p.s.k. signals are given in Figure E.1 and Table E.1 from which:

$$\text{bandwidth/channel} = 1{\cdot}25 \times \text{symbol rate}$$
$$= 40 \text{ KHz}$$

Allowing for guard bands, the channel separation is 45 kHz. Hence the transponder capacity is

$$\frac{36 \times 10^3}{45} = 800 \text{ channels}$$

(b) *P.c.m./p.s.k./t.d.m.a. system* Again 10 per cent of the noise is usually allocated to interference and the same amount to the up-path. The operating C/N is then

$$\frac{C}{N_T} = E_S - P_L + \frac{G}{T} - 10 \log(kB) - 10 \log \left(\frac{10}{8}\right) - M \qquad (M = 3 \text{ dB})$$

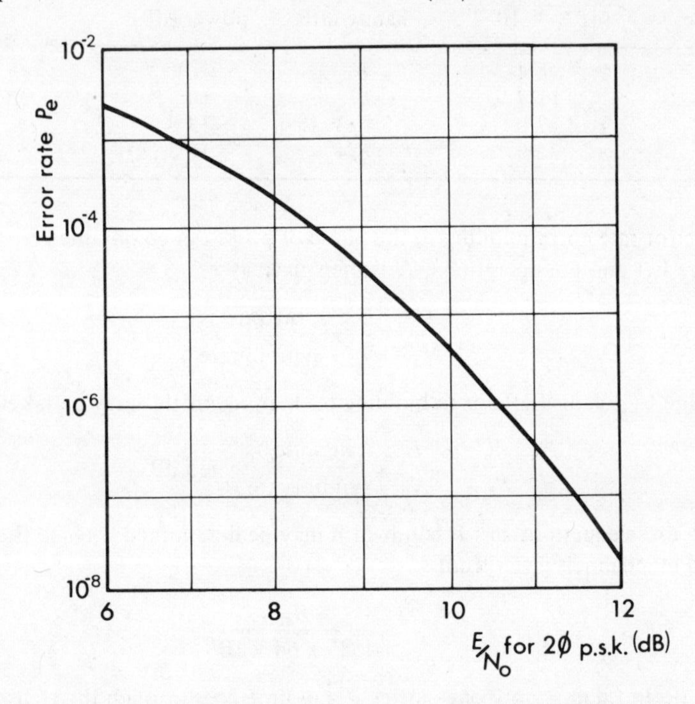

Figure E.1 Error rate of coherent p.s.k. systems. For 4-phase add 3 dB to E/N_0 and double P_e. For 8-phase add 8·8 dB to E/N_0 and double P_e. For 16-phase add 14·8 dB to E/N_0 and double P_e

Table E.1.

P.c.m. channel characteristics

8 kHz sampling frequency
8 bits/sample
64 kbits/s per channel
'A' law companding

P.s.k. modulation–coherent detection

Error rate for 2-phase $P_e = 0{\cdot}5(1\text{-}E_{\text{rf}}\sqrt{(E/N_0)})$

For m-phase, error rate at $E/N_0 = x$ is twice that of a 2-phase system with

$E/N_0 = x \sin(\pi/m)$.

i.e.
$$P_e = 1 - E_{\text{rf}}\left[\sin\frac{\pi}{m}\sqrt{\left(\frac{E}{N_0}\log_2 m\right)}\right] \qquad m > 2$$

Bandwidth theoretical minimum = bit rate/$\log_2 m$

Practical bandwidth = 1·25 x theoretical bandwidth

m phases	Theoretical C/N for $P_e = 10^{-4}$	Theoretical bandwidth	Relative power (dB)	Practical C/N rel. to theoretical (dB)
2	8·4	b	0	+1·5
4	11·7	$b/2$	+0·3	+1·5
8	17·0	$b/3$	+3·8	?
16	22·9	$b/4$	+8·5	?

The error-rate corresponding to the phase of p.s.k. can be obtained from Figure E.1 and the operating C/N is then given as

$$\frac{C}{N_T} = \frac{E}{N_0} \times \frac{\text{bit rate}}{\text{symbol rate}}$$

but due to practical effects in building p.s.k. modems this may be taken as

$$\frac{C}{N_T} = \frac{E}{N_0} \times \frac{\text{bit rate}}{\text{symbol rate}} + 1{\cdot}5\text{ dB}$$

From these equations an r.f. bandwidth may be determined whence the number of t.d.m. channels given from

$$\frac{B_{\text{rf}}}{1{\cdot}25 \times 64 \times 10^3}$$

N.B. In t.d.m.a. only one carrier at any time goes through the transponder; hence no back-off is needed and no intermodulation results. (The limit to capacity will, however, be intersymbol interference).

For multiple carriers the equations are the same as for a single carrier except that guard bands and pre-amble etc. reduce the time efficiency; in the MAT 1 system by only 5 per cent. Whence the number of channels–single channel per carrier x 0·95.

In performing the down-link equation note that the transmitter power is only required in bursts, so large powers may be calculated but the duty cycle must be taken into consideration.

Index

Index

Index